ESSAY ON
THE THEORY OF THE EARTH:
ELECTROMAGNETISM IN UFOS AND THE ORIGIN OF MASS EXTINCTIONS AND THE ICE AGES

ROBERT ITURRALDE

A DIVISION OF HAY HOUSE

Copyright © 2021 Robert Iturralde.

All rights reserved. No part of this book may be used or reproduced by any means, graphic, electronic, or mechanical, including photocopying, recording, taping or by any information storage retrieval system without the written permission of the author except in the case of brief quotations embodied in critical articles and reviews.

Balboa Press books may be ordered through booksellers or by contacting:

Balboa Press
A Division of Hay House
1663 Liberty Drive
Bloomington, IN 47403
www.balboapress.com
844-682-1282

Because of the dynamic nature of the Internet, any web addresses or links contained in this book may have changed since publication and may no longer be valid. The views expressed in this work are solely those of the author and do not necessarily reflect the views of the publisher, and the publisher hereby disclaims any responsibility for them.

The author of this book does not dispense medical advice or prescribe the use of any technique as a form of treatment for physical, emotional, or medical problems without the advice of a physician, either directly or indirectly. The intent of the author is only to offer information of a general nature to help you in your quest for emotional and spiritual well-being. In the event you use any of the information in this book for yourself, which is your constitutional right, the author and the publisher assume no responsibility for your actions.

Any people depicted in stock imagery provided by Getty Images are models, and such images are being used for illustrative purposes only.
Certain stock imagery © Getty Images.

Print information available on the last page.

ISBN: 978-1-9822-7454-2 (sc)
ISBN: 978-1-9822-7455-9 (e)

Balboa Press rev. date: 09/16/2021

Contents

Introduction ix

1 Electromagnetism in UFOs and Its Effect on Planet Earth 1
2 The Biblical Flood and the UFO Phenomenon 15
3 The Origin of Boulders, Nature's Puzzles, and the
  UFO Phenomenon 43
4 UFOs, Asteroids, Comets, and Mass Extinctions 67
5 The Strange Death of the Megafauna and Flora
  during the last Pleistocene Mass Extinction 93
6 The Origin of the Ice Ages and the UFO Phenomenon 145
7 The Tragedy of the Human Condition 156

Conclusion 161
Bibliography 163
About the Author 167

To my parents because they taught me the love for books.

Introduction

An intelligence knowing at a given instance of time, all forces acting in nature, as well as the momentary position of all things of which the Universe consists would be able to comprehend the motions of the largest bodies of the world and those of the lightest atoms in one single formula, provided his intellect were sufficiently powerful to subject all data to analysis; to him nothing would be uncertain but past and future would be present in his eyes" (Pierre Simon de Laplace).

I will say with confidence that all these attributes are in the intelligence behind the UFO phenomenon. Two of the greatest mysteries in human civilization are, in paleontology, the cause of mass extinctions, and in geology, the causes of the Ice Ages. Interestingly, the great Charles Darwin was very impressed by his observation in South America of millions of fossils of extinct animals. A megafauna and flora never seen before, with species that apparently had all the natural resources to live a long natural life. Shockingly, all died in their prime and suddenly. According to geologists, their end came in a recent geological age.

Darwin wrote in his January 1834 journal about his travel to South America, "It is impossible to reflect on the changed state of the American continent, without the deepest astonishment, formerly it must have swarmed with great monsters, now we find mere pigmies, compared with the antecedent allied races. The great number, if not all of these extinct quadrupeds lived at a late period and were the contemporaries of most of the existing seashells. Since they lived, not very great change in the form of the land can have taken place. What, then has exterminated so many species and whole genera; the mind at first is irresistibly into the belief of some great catastrophe; but thus to destroy animals, both large and small in Southern Patagonia in Brazil, on the Cordillera of Peru, in North America up to Behring's straits. We must shake the entire frame work of the globe, no lesser physical event could have brought about this wholesale destruction not in the Americas, but in the entire world."

Darwin proceeded, "It could hardly have been a change of temperature, which at about the same time destroyed the inhabitants of tropical, temperate and arctic latitudes on both sides of the globe. Certainly, it could have been man in the role of destroyer and were he to attack large animals. Would he also be the cause of extinction of the horse? Did those plains fail of pasture, which have since been overrun by thousands of and hundreds of thousands of the descendants of the stock introduced by the Saniards? Certainly, no fact in the long history of the world is so startling as the wide and repeated extermination of its inhabitants."

My theory is that the intelligence behind the UFO phenomenon caused mass extinctions and the Ice Ages. After researching UFOs for twenty-six years, my conclusions are only circumstantial.

CHAPTER 1

Electromagnetism in UFOs and Its Effect on Planet Earth

Electromagnetism in UFOs

> The mechanisms behind the magnetic field and behind the reversals are still mysterious . . . one of the grand intellectual challenges . . . in all of the physical sciences.
> —Raymond Jeauloz

It is well known since the 1940s that UFOs caused electrical disturbances in cars, household appliances, and compasses, and blacked out power stations and jammed radars. In an interesting case, on February 9, 1962, a passing UFO made a car lose power, but the lights were unaffected. Also, there are cases in which only a car's motor or headlights are affected.

We see in this type of incident intelligent selectivity, which shows that UFOs are intelligently controlled. Any natural force field would act without directedness and always black out everything within range. Surely only beans of very well-controlled energy, by the proper selection of interfering waves, can blink out a light and leave the radio working. Furthermore, in this example, on August 17, 1959, in the city of Uberlandia, Minas Gerais, Brazil, a round UFO was reported by witnesses, following a power trunk line, after the keys of the power station were turned off. As a matter of fact, whole towns and the entire region were left in darkness. This celebrated case caused hysteria and a media frenzy. For a short period, the region was deprived of electricity; after the UFO left, the keys turned back on by themselves and there was no damage to the electrical circuits.

Similarly, in April 1952, there was a sudden blackout of a power system in northeastern Ohio, and two days later, an unexplainable blackout at a power station in Evaston, Indiana, which lasted for two hours. The police and the public were puzzled that at the same time there were UFO sightings. Indeed, 1965 was a record year for UFO sightings. In November 9 of the same year, there was a big blackout. A city aviation commissioner of Syracuse, New York, was flying a plane with other individuals when the lights went out at 5:22 p.m. He managed to land safely but reported a UFO: "A ball of fire south of us, toward Thompson road and carrier traffic circle, appeared to be about 100 feet in the air and 50 feet in diameter."

The blackout plunged eighty thousand square miles and thirty million people in seven states and parts of Canada in the dark. Certainly, there were many eyewitnesses near Syracuse that reported a ball of fire before the lights went out.

Likewise, another eyewitness was preparing to land his private plane at Hanck Field when he saw a UFO about one hundred feet in diameter near the New York power company, which then passed over the New York Central Railroad tracks. Similarly, a flight instructor reported a UFO: "A ball of fire of orange-red which flared up, bigger than a house, hovered over the high line which runs from Clay to Niagara Falls." Moreover, two witnesses in Somerville, New Jersey, reported seeing a UFO, "A very large light, larger than the evening star." UFO researchers have collected more than one hundred cases of blackouts caused by UFOs. My question is, how does the electromagnetism in UFOs affect Planet Earth? How did the electromagnetism in UFOs create the Biblical Flood? How did Essay on the Theory of the Earth: Electromagnetism in UFOs and the Origin of Mass Extinctions and the Ice Ages13 13 the electromagnetism in UFOs teleport the great boulders around the world?

Earth's Magnetism

The sun and the outer four planets, Jupiter, Saturn, Neptune, and Uranus, have magnetic fields. In fact, Earth is the only inner planet with a strong magnetic field because Venus, Mars, and Mercury don't have magnetic fields. Although, the solar system was formed 4.5 billion years ago, and the interior of the inner planets is similar. They are formed with a dense metallic core, a less dense mantle, and a rigid crust. So my question is, why is only Earth magnetic? And how and why are electric currents generated inside the earth? Interestingly, some scientists have suggested that the heat of the sun is somehow transformed into electrical energy, although they can't name the mechanism. Also, some scientists believed that the gravitational and geomagnetic fields involved an action at a distance that originated within the earth and created the electrical currents inside the earth.

A scientist named Gauss came with a theory called spherical harmonic analysis, which he thought would provide a neat and elegant mathematical way of describing earth's magnetic field. Nevertheless, it didn't solve the mystery of what actually caused the origin of the magnetic fields. The idea is to find what the internal sources are and any possible external source of the origin of the magnetism on Planet Earth.

Gauss was sure that the primary source of earth's magnetism was internal. On the other hand, another scientist named Orsted discovered that a magnetic field could be the result of electrical currents. Likewise, Gauss believed in the idea of permanent magnetism although he didn't like the idea of permanent dipoles deep within the earth, revolving or processing fast enough to produce the observed rates of change of declination and inclination and now intensity of the magnetic field. He suggested instead that iron particles in the crust were the source of earth's magnetic field. Further, in accordance with modern notions of a fluid-filled earth, he thought that the solid crust must be relatively thin but that it was gradually thickening as the fluid beneath solidified on to it "becoming magnetized in the process and so causing local changes in the direction and intensity of the magnetic field." Actually, he supported this theory with the observation that the geomagnetic intensity was greatest at the poles where the temperature was lowest and the magnetized crust was expected to be thickest.

By the same token, Gary Glatzmaier and Paul Robert's computer model had demonstrated that a magnetohydrodynamic dynamo that obeyed Maxwell's equations of electromagnetism and Stoke's equations of fluid dynamics could generate an earthlike magnetic field. The result of their research is that the magnetic field, after the equivalent of 300 years, it had still not achieved another polarity reversal. It has been 780.0008 years since the last documented reversal although paleomagnetism has told us that the past is not an indication of the future.

According to modern science, polarity reversals occurrandomly, and the intervals between them range from a few thousand to tens of millions of years, although examination of paleomagnetic records has revealed that there had been several major excursions of the magnetic fields' direction, much bigger than regular secular variation and not eventuating in full polarity reversals. For instance, there are the so-called Laschamp excursion recorded in forty-thousand-year-old lava flows in the Chaine Des Puys, region of France and the Mono Lake excursion, thought to have occurred in California about twenty five thousand years ago. A question remains and this is, which is the source of the energy required to drive the dynamo to reversal? Moreover, convection in the outer core, where the fluid is thought to flow as easily as water, is up to a million times faster—millimeters per second, or tens of kilometers per year. My question is, what this motion? Certainly there is one obvious source, and that is the heat accumulated in the core when the earth first formed.

According to scientists, a large-enough temperature difference between the inner core and the core-mantle boundary would drive thermal convection. Heat supplied by decay of radioactive element in the core would aid the process, but it is uncertain whether a sufficient source of radioactive heat exists within the core. My other question is, how has the radioactive heat lasted for 4.5 billion years! Another unsolved mystery is, what actually happens during a polarity reversal? Computer simulations and paleomagnetic records agree that the field intensity is reduced considerably. As a matter of fact, mass extinctions—the most well-known is the extinction of the dinosaurs sixty-five million years ago, at the famous Cretaceous-Tertiary boundary—have not occurred close enough to magnetic reversals for us to claim cause and effect. In fact, in the computer models, there are polarity reversals at random intervals of time. Nevertheless, my question is, what is the physical trigger that will initiate the necessary changes in core fluid motion and make the electromagnetic fields reverse?

After all, direct proof of the geodynamo remains an elusive science fiction. In addition, in the past two hundred years, the strength of the magnetic field has dropped by some 15 percent; this rate of change has astonished many scientists. Further, scientists believed that at this rate of change, we could head for the next polarity reversal. Polarity reversals occurred in the early Quaternary period, about a million years ago. Furthermore, some scientists believed that magnetic reversals occur roughly every one million years. In fact, some scientists believed that the earth's magnetic field had flipped on average every 430 years.

The mechanisms behind the magnetic reversals are still really mysterious, one of the great intellectual mysteries in all of the physical sciences. My question is, how might electrical currents be generated inside the earth? Further, the late great scholar Immanuel Velikovsky postulated the theory that the earth's magnetism can be reversed by powerful cosmological lightning strokes. According to Velikovsky, the last magnetic reversal took place in the eighth century, before our present era or twenty-seven centuries ago! The observations were made on clay fired in kilns by the Etruscans and Greeks. On the other hand, orthodox geology and science claimed that the last magnetic reversal took place 780 years ago! Nevertheless, many scientists and scholars have found many magnetic reversals in historical times. Unfortunately, orthodox science and geology do not mention those findings.

For instance, in 1896, Giuseppe Folgh began his study of Attic, Greek, and Etruscan vases of different centuries, starting with the eighth century BPE. His conclusion was that in the eighth century, the earth's magnetic field was reversed in Italy and Greece. Italy was closer to the south than the north magnetic pole. Moreover, a famous geologist and scholar named P. L. Mercanton of Geneva claimed that the magnetic field of the earth was disturbed some time, during, or immediately following the eighth century "to the extent of complete reversal." Similarly, another geologist named Manley speaks of the earth's magnetic field reversal in historic times, 2,500 years ago! In addition, there is evidence of a magnetic reversal 3,500 years ago. By the same token, two scientists, J. P. Kenneth and N. D. Watkins reported that geomagnetic field reversals have occurred more than twenty times during the past four million years, and probably more than one hundred times during the tertiary period.

Physicist James M. McCanney stated, "The magnetic fields of Jupiter and Saturn discovered by Pioneer [a rocket sent to explore the solar system] have created an unmentionable dilemma in the astrophysics community. In short magnetic fields do not self-generate and sustain themselves for billions of years!"

Actually, it seems that Maxwell's equations on electromagnetism have been swept under the theoretical rug by some traditional scientists who believed that the only force acting in the cosmos is gravity. Interestingly, for magnetic fields to form, electric currents must flow. A potential difference must be maintained because not only does the earth magnetic field shift its direction, it also changes in strength. The origin of the earth's shifting, twisting, and flipping of the magnetic field is still a mystery for the science community. My question is, what causes the electrical currents in the earth's core?

The earth's magnetic field can change erratically in direction and in strength or even flip over altogether. Obviously, the problem is so complicated that scientists cannot predict how the magnetic field will behave in the future or why it has changed at a particular time in the past. Further, what the late scholar Immanuel Velikovsky thought was Venus that caused the electromagnetic field to reverse, in my opinion, were UFOs. Similarly, what the ancient Greeks, Romans, and the ancient civilizations thought was the star Phaeton performing acrobatics were also UFOs. Moreover, UFOs are responsible for reversing the magnetic fields. Scientists still ask what triggers the reversing of the magnetic fields. Interestingly, some scientists postulate a large cosmic body, probably comets, asteroids, or even a planetoid. For instance, the late great scholar Immanuel Velikovsky postulates the planet Venus as responsible for reversing the magnetic fields of the earth.

In my opinion, UFOs are responsible for the reversing of the magnetic fields and mass extinctions and the Ice Ages. Further, whatever the cosmic body is, electrical effects are more possibly responsible for the electromagnetic reversals. Moreover, if the most likely scenario is a cosmic body in electromagnetic contact with the earth, they will be debris left in the form of tektites. In fact, have been artificially produced by subjecting earth materials to a beam of electrons and also by immersing cold glass spheres in the plasma jet from an electrical arc. Interestingly, some ancient texts mention that tektites or fire pearls fell from the sky. Furthermore, scientists do not believe that an actual impact could be the trigger of a magnetic reversal, but a near miss by a cosmic body one thousand kilometers in diameter would have a considerable gravitational and electrical effect. Moreover, the scientists asked about what could be the possible trigger of magnetic reversals. Scientists said, "We couldn't rely on some supernatural explanation, like something happening in the heavens of a vague character, which actually violated the laws of nature; or it had to be something natural. A part of nature's ordinary structure, which disrupts the earth's inner electrical and magnetic structure, whenever it happens."

As a matter of fact, there is evidence that there must have been a relatively recent unknown catastrophic event. This event imparted magnetic changes to igneous rocks ten to hundreds of times stronger.

Undoubtedly, this cosmic body is what ancient civilizations called Phaeton visitations. Indeed, it is a coincidence that a well-marked reversal took place around 11,500 years BP. Although there is no evidence of a cosmic body visiting the earth, there is a vast evidence of circumstantial evidence that UFOs were already visiting Planet Earth. Scientists explain that the earth, being a rotating, approximately spherical, electrically magnetic body, is like a giant magnet. The earth possesses an electrically-charged magnetic field called geomagnetism. Likewise, laboratory experiments have shown that a rotating spherical magnet takes with it a magnetic induction field, which in turn gives rise to the induced electromagnetic force.

Indeed, only UFOs are the best candidates to be the cause of the last Pleistocene mass extinction and the Ice Ages. We may therefore dismiss the theories, which postulated an asteroid, a giant meteor, or a comet. In fact, it is wrong to ascribe the last Pleistocene mass extinction to a comet, since comets lack the mass to cause the earth changes we see in the Pleistocene mass extinction. Similarly, the late great scholar Immanuel Velikovsky advocated the planet Venus as the cause of the last Pleistocene mass extinction.

He identified planet Venus after having eliminated meteors, comets, and asteroids as possible candidates, although there is no evidence for any planet having ever approached Earth close enough to affect its rotation or axial tilt. Or that Mars or Venus ever caused the widespread mass extinctions known to humankind. Moreover, many astronomers have stated that the available observational data suggest strongly that no planet-sized object has passed near earth in historic times and probably not for a much longer time.

Nonetheless, something capable of exerting a tremendous influence approached earth closely enough to have caused mass extinctions and the Ice Ages in 11,500 BP. The effects left on Planet Earth needs something of a planetoid dimension. Something close to earth's mass and density or of smaller mass and greater density like UFOs. There is, however, no evidence for definitely identifying the astronomical visitor.

Also, some scientists postulate a supernova fragment that entered our solar system and caused mass extinctions. Although it is impossible at the present time to know the nature of the celestial body, that ancient civilizations called it Phaeton. Obviously, from the descriptions given by ancient civilizations, it seems to me more like UFOs than stars, asteroids, or comets. The earth is a giant magnet, and an electromagnetic object like a UFO can affect the earth's magnetic fields. UFOs can produce a thermal effect and shift the terrestrial axis, even change the rotational velocity of the earth. Moreover, this in turn would have a thermal effect, because the energy of motion would be converted into heat and other possible forms of energy like electrical, magnetic, or nuclear. Moreover, an electromagnetic and radioactive field would produce the phenomenon associated with a glacial period like extreme heat and cold.

The great scholar George Gamow wrote in his masterpiece, A Biography of the Earth, "However, up to the present time, we still do not know what causes this magnetic field and according to our best knowledge of the properties of the earth's interior it should not be there at all!" As a matter of fact, investigation of the magnetic properties of different substances such as iron and nickel proves definitely that any trace of magnetization must completely disappears soon as these substances are heated above the so-called Curie Point since the temperature inside the earth reaches values much above the Curie Point. Certainly, one can hardly expect that the observed phenomena can be explained as the result of permanent magnetization. Further, the most natural hypothesis, according to which the source of terrestrial magnetism is situated in the central iron core, can hardly stand up because seismological evidence seems to show that this iron is completely molten. Still, under very high pressures that turn melted rocks into a plastic mass, the magnetic properties of iron and other materials can be modified at even at much higher temperatures. For example, in a laboratory, a compressing machine, which permits the study of the properties of matter at pressures up to 220,000 atmospheres or equivalent to the pressure at a depth of 480 kilometers under the surface of the earth. We get in this experiment the indication if materials at great depths do possess magnetic properties. The question still remains of the origin of the magnetization. Another hypothesis proposed for the explanation of terrestrial magnetism is that the earth is not a permanent magnet but rather an electromagnet created by unknown forces, and that the earth is being fed by some electric currents now flowing the earth's body. Obviously, the problem arises when we ask about the origin of these electrical currents. However, all attempts to answer this question find no answers. So now in the twenty-first century, we still don't know why the magnetic needle points north.

Convection Currents

Another theory about the origin of the terrestrial magnetism is that convection currents deep in the interior of the earth produce the electromagnetic fields. According to this theory, the convection currents produce an uneven heating of the crust, which cause thermoelectric currents to flow along the equator. Although, other experiments have shown that the convection currents are too slow to produce the magnetic fields. Moreover, what still remains a mystery is the nature of the geodynamo that is supposed to produce the magnetic fields, postulated by some geologists. It is clear that it has been active for more than 4 billion years! Further, it seems to have the same strength since the creation of the planet! Likewise, other geologists think that magnetism is an innate property of the liquid outer core. In fact, the metallic core is a very good conductor of electricity and also is a fluid, capable of movement. So they think a magnetic field must be generated by the interaction of these two components of the earth.

Certainly, I think that UFOs charge electrically the convection currents and keep the geodynamo and the magnetic fields active. In fact, an electric current produces a measurable magnetic effect. The UFO phenomenon produced the action at a distance of invisible vibrations in the inner core and plasma. So UFOs projected electrical currents in the core, and plasma generated the magnetic fields. As a matter of fact, there must be energy available to drive the fluid motion, and the

UFO phenomenon is responsible. Moreover, the UFO phenomenon is responsible for the magnetic field reversals and the mass extinctions, the Flood, and the Ice Ages.

Some geologists postulate the geodynamo as the force responsible for the electromagnetic fields. My question is, where does the energy that drives the geodynamo come from?

And what is the nature of the energy source, which must be constant for 4'000.0000.0000 billion years?

Scientists believed that thermal convection in the outer core is one possibility. A kind of deep, simmering turnover of the molten layer providing the motor for magnetism. On the other hand, in the 1980s during the UFO wave in Belgium and most recently in 2002 with very sophisticated technology, UFO photographs were analyzed. The scientists discovered through computer analysis of the photographs that it revealed a halo of something lighter surrounding the craft. Further, especial optical processing shows that within the halo, the light particles form a certain pattern around the craft, like snowflakes in turbulence. Interestingly, this is very similar to the pattern of iron fillings, which is caused by "the lines of force" in the magnetic field. In fact, this could indicate that the craft is moving by using a magnetoplasmadynamic propulsion system. Furthermore, some geologists believed that during the Permian Era, 225 to 280 million years ago, the magnetic fields remained reversed for at least fifty million years. Conversely, some geologists postulated that the average longtime scale for magnetic reversals varies between forty and eighty million years. Likewise, there were only forty reversals, one per million years on average, although in the past forty million years there have been 143 reversals, more than three times as many.

By the same token, scientists in a computer simulation produced the most shocking results. The results showed that after the equivalent of forty thousand years, the outer core field destabilized and waited long enough for the inner core field to regenerate in the original direction. However, the magnet flipped and grew back in the opposite polarity. The computer model had undergone a full geomagnetic polarity reversal.

The computer simulation had maintained a stable dipole field for forty thousand years. The computer model seemed to have tried to reverse several times but had almost always reverted to its original polarity, and eventually, just once, it succeeded in full reversing of the magnetic fields. Indeed, the computer simulation had answered Einstein's ninety-year-old question.

Further, in important scientific papers, like the journal "Nature and Physics of the Earth," the researchers convince the world of science that electric currents in the earth's molten, iron-rich outer core, brought about by the combined effects of convection and the planet's rotation, were all that was needed to account for everything known about earth's magnetic field. Moreover, the late great scholar Immanuel Velikovsky said, "A thunderbolt when striking a magnet reverses the poles of the magnet." Similarly, the intelligence behind the UFO phenomenon strikes the magnetic fields, causing the reversal.

For example, as they cool off from a molten state, it will allow the iron particles that they contain to became aligned with the earth's magnetic field. The magnetism is called remanent magnetism; even a sedimentary such as sandstone, if heated and cooled, slowly does this.

Interestingly, according to paleomagnetic studies, the direction of the magnetic field will change if the rock is struck by a thunderbolt! Also, lightning changes the magnetic direction in rocks! Indeed, this is the form the intelligence behind the UFO phenomenon strikes the magnetic fields and reverses it. Furthermore, in Science News, vol.125, no. 24, June 16, 1984, researchers Downey and Tarling from the University of Newcastle in England wrote that the destruction of Late Minoan sites on Crete occurred in two events separated by as much as thirty years. The researchers found that the magnetic signatures at archaeological sites on eastern Crete were identical to one another. Moreover, the sites in Eastern Crete matched those of the deposits that followed the volcano's tremendous explosion.

According to geologists, during the second phase, the volcano collapsed forming the huge caldera. This evidence corroborates what was found in Loch Lomond and Lake Windermere of geomagnetic changes around 3,500 years ago. The UFO phenomenon, with a beam of lightning, strikes the magnetic fields and convects currents, causing the magnetic reversals.

Further, it is widely accepted that electric currents are a major component of the geomagnetic field, with the dynamo being driven by the combined gravitational influence of the sun and moon. The cause of precession in the first place, thus any serious distortion (up or down) of this solar/lunar gravitational pull, would disrupt the precession so that the earth's magnetic field would drop. Geologists agreed that the earth's electromagnetism is generated in the earth's core and emanates from the movements of the molten fluid. As a result, this motion is created by the effects of the earth's precession on the solid crust and on the core itself. As a matter of fact, my theory is that the intelligence behind the UFO phenomenon charges the molten fluid with electromagnetism, reversing the magnetic poles and causing mass extinctions, Ice Ages, and disasters like the flood mentioned in the Bible.

Furthermore, some paleomagnetologists stated that a particularly well-marked reversal took place around the general date, 11,500 BP. Likewise, reversedy magnetized sediments near Gothenburg in southern Sweden have been dated as being between 12,600 and 8,600 years old. Interestingly, according to paleomagnetologists, the average of these dates came close to 11,500 BP, also within the so-called Laschamp geomagnetic event that took place 17,000 to 7,000 years ago. Additionally, strong paleomagnetic anomalies have been discovered in North and Central Europe, Eastern Canada, the Gulf of Mexico, and New Zealand, around 12,500 BP. Some scientists regarded as representing a global magnetic reversal about that date, which has come to be named the Gothenburg Flip. Moreover, a drop in the strength of the earth's entire magnetic field appears to have occurred between 12,350 years and 13,750 BP. Also, these magnetic reversals were accompanied by earthquakes, volcanisms, water table fluctuations, and large-scale climatic changes.

For example, the great Biblical Flood was caused by UFOs interfering with the magnetic fields. Similarly, according to scientific research of eight observed micro-faunal extinctions in the Southern Ocean, six either occurred during or very close to geomagnetic polarity reversals. By the same token, during the last Pleistocene mass extinction, it took place in the uplift of the Rockies mountains, the Alps, the Himalayas, and other large mountain ranges around the world. These mountain ranges attained these present elevations at the close of the so-called Pleistocene times.

A great scholar, R. T. Chamberlin, wrote, "Hundreds, if not thousands of cubic miles of the body of the earth almost instantaneously heaved upward, produced a violent earthquake which spread throughout the entire globe. Many world shaking earthquakes must have been byproducts of the rise of the sierra."

Another great scholar, Edward Suess, wrote, "The earthquakes of the present day are certainly but faint reminiscences of those telluric movements to which the structure of almost every mountain range, bears witness. Numerous examples of great mountains chains suggest by their structure . . . episodal disturbances of such indescribable and overpowering violence, that the imagination refuses to follow the understanding."

Chapter 2

The Biblical Flood and the UFO Phenomenon

> Evidence of massive flooding and drainage of the earth's land masses during geologic history can be explained by the Dynamic Axis Theory, using small movements of the earth's axis of rotation. The changes in sea level around the world can vary from +1,238 feet to -1238 feet for each degree of axis shift.
>
> —Mac B. Strain

According to the great late historian Will Durant, there are about eighty thousand myths, legends, traditions, and religions that mention a great flood in historical times. Moreover, there are thousands of myths and legends that mention strange phenomena in the sky associated with natural disasters. For example, the Incas of Peru mention that a horrible flaming phenomenon occurred in the skies, and they thought that it was their god and called it Quetzalcoatl. Likewise, the Babylonian myth narrates that Marduk with four assistants wandered in the skies. Marduk was also called Phaeton by the Greeks. Furthermore, the Vedic mythology of ancient India describes Indra (Phaeton) as traveling across the skies, accompanied by a band of destructive entities known as Maruts who are described as "blazing, brilliant with fire shining like snakes with long trailing tails. The terrible ones and the shakers of heaven and earth."

As we see in the twenty-first century, we call these objects flying in the skies UFOs. For instance, some myths mention several fiery or luminous objects that arrived simultaneously in the neighborhood of earth. Moreover, Phaeton is described by ancient writers like Ovidio and Euripides as "a bright, fiery body." Interestingly, Phaeton signifies "the shining one" and the Roman writer Pliny "the Old" describes Phaeton as follows: "It had a fiery appearance and was twisted like a coil, and it was very grim to behold; It was not really a star so much, as what might be called a ball of fire." Furthermore, the seventeenth-century writer named Adam Rockenbach who claimed to have read the most famous ancient writers wrote about Phaeton: "It was fiery, of irregular, circular form, with wrapped head, it was in the shape of globe and was of terrible aspect in the form of a disc." Similarly, an ancient Jewish tradition mentioned that "the deluge was caused, by the lord god changing the places of two stars in a constellation."

So Phaeton was identified as a fiery star-like body. Certainly, the ancient Hindu myth of Brahma is described as arriving in the heavens: "An exceedingly small white boar-shaped which in the space of an hour, grew to the size of an elephant of the largest size, and remained in the air. Prior to it plunging to earth and causing a worldwide flood." Similarly, Chinese traditions mention that during the time of the legendary emperor Ya-Hou, "a brilliant star issued from the constellation of Yin, prior to a tremendous upheaval." The object is described like a star or sun rather than a comet; "It roamed the skies at will and sometimes came close to earth in its heavenly wandering." Similarly, the Incas observations of "unusually moving stars" allowed the Inca deluge hero to find a mountain refuge prior to the advent of a world-drowning flood. Moreover, the ancient world believes Phaeton as a "round, brilliantly, fiery body of appreciable size and much more sun and star like, than a comet." As a matter of fact, the belief among the ancients was that Phaeton caused the famous deluge mentioned in the Bible like the Flood.

Interestingly, many ancient legends and myths mention two or more cosmic visitors punishing the earth simultaneously. For example, the Norse myth mentions two rampaging celestial monsters, the Fenris Wolf and the Midgard Serpent that brought death and ruin to the ancient world. The celestial phenomenon is described as "speeding through the heavens, side by side, they acted in concert." Further, old Persian texts mention a similar pair of ravaging celestial bodies that brought destruction to the earth. In the Persian mythology, two celestial bodies called Zohak and Iblis brought disaster and suffering to the inhabitants of Earth.

Witnesses reported a serpent formed like a dragon. Ancient civilizations described these objects as frightening dragons or coiling serpents. Interestingly, sometimes they were described as multiheaded with wide, gaping jaws, tusks, or horns; sparkling crowns or growing manes and hairlike appendages; speckled bodies; and jetted great streams of fire fr-om their mouths. Similarly, other legends and myths mention what they thought was Phaeton—or whatever cosmic body it was—emitting pestilential clouds, poisonous blasts, rains of red milk, or blood. Likewise, other legends mention that the star-like body called Phaeton produced hissing, roaring, thunderous, or explosive sounds. Certainly, asteroids or meteorites do not produce these types of sounds. According to these reports, these objects had bulk and mass. Also, Phaeton was reportedly so brilliant like the sun.

Indeed, what ancient civilizations called the celestial body Phaeton was none other than a UFO or UFOs. For example, the descriptions of UFOs in the nineteenth, twentieth, and twenty-first century are similar to the descriptions given by ancients of the celestial phenomenon that they called Phaeton.

Actually, an ancient document from the times of the Egyptian king Thutmose III, who died about 1450 BC, was the Tulli Papyrus. In this report, UFOs are mentioned. "In the year 22, third month of winter. The scribes of the house of life found it was a circle of fire that was coming in the sky. They went to the king to report it. Now after some days had passed, Lo! they were more numerous than anything, and they were shining in the sky. The army of the king looked on, then, upon the circles went up higher and to the south. It was a marvel never occurred since in this land."

Moreover, the great writers of antiquity like Pliny the Older reported UFOs in the only form that their cultural miliun can describe. "Several suns were seen at midday at the Bosphorus and this lasted from dawn to sunrise. At Praeneste a torch was seen in the sky. In 93 BC, at Volsinii, flame seemed to flash from the sky at dawn. After it had gathered together the flame, the flame displayed a dark grey opening and the sky appeared to divide. In the gaps tongues of flame appeared." Similarly, in 91 BC, "at sunset a globe of fire in the northern region rushed across the sky, emitting tremendous sound." Likewise, in Spoletium, "A gold colored fireball rolled down to the ground and, growing larger, rose from the earth towards the east becoming large enough to blot out the sun."

As we see, all these descriptions given to UFOs in antiquity are trying to describe something unusual in culturally familiar terms. So what ancients described are what people in the twentieth and twenty-first century call UFOs. Additionally, ancient myths describe Phaeton (a UFO) discharging inflammable hydrocarbons and methane, which is exactly what the La Brea Tar Pits in California are made of as a result of the electromagnetic exchange between UFOs and the magnetic fields of the earth. By the same token, ancient myths and traditions describe immense inflammable clouds, aerial explosions, and streams of fire directed upon earth by celestial objects that now look like UFOs. Additionally, ancient myths mention calamities brought to earth by Phaeton (a UFO) in the form of sticky or inflammable bloodlike fluids falling from the sky (La Brea Pits in California is made of a sticky substance) and hot naphtha, bitumen, or rains of fire. A similar material was found in the Pits of La Brea where millions of animals have been found. The possible hydrocarbon origins of some of the material, also the fluids, are described as bloodlike as they consisted of a brownish red liquid. In one myth from Finland, this brownish liquid is compared to red milk. Obviously, this description suggests that this substance was thick and opaque. Like the sticky and inflammable substances, these red rains were destructive to life in the planet. Myths mention a stream of blood which gushed out on the mountain when it was hit by a thunderbolt. Similarly, the apocalypse of Thomas mentions that a "great cloud of blood shall come down . . . a rain of blood upon all the earth." Likewise, the Maya legend mentions a time of tremendous world upheaval, when the waters of all the rivers turned to a bloody redness. Many legends and myths mention that this red liquid arrived and fell as a great cloud, as a stream, and that it rained all over the world.

It seems to me that vast quantities of this substance fell all over the world. My question is, what was its origin and its true nature? Further, ancient Mesopotamian texts mention the apparition of a celestial object they called Phaeton that brought huge tectonic upheavals on the planet. Similarly, the Roman writer Ovid mentions a celestial body that ancient civilizations called Phaeton. The approaching of this celestial object caused the following: "The earth burst into flames, the highest parts first, and splits into deep cracks, and its moisture is all dried up. The meadows are burned to white ashes, the trees are consumed, green leaves and all, and the ripe grain furnishes fuel for its own destruction. Great cities with their walls and vast conflagration reduced whole nations to ashes."

Interestingly, in the twentieth century on August 7, 1970, a red glowing ball swept through the village of Saldene in Ethiopia. It destroyed houses, uprooted trees, melted asphalt on the road, burned grass, and broke the stone wall of a bridge into pieces. The UFO made an earsplitting sound then became stationary. About fifty houses were damaged, eight people were injured, and one little girl died. By the same token, on July 31, 1967, in Caracas, Venezuela, just before an earthquake struck, many witnesses saw a huge, glowing red ball swept across the sky. Additionally, on September 15, 1749, a glowing object or UFO roared and whirled in the countryside in England, frightening cattle. Also a witness saw the UFO sucking water from a pond before roaring away, splitting and smashing trees and seeming to dart arrows of light into the ground. Furthermore, in a classical Greek work, Theogony clearly described a "great conflagration which preceding the Flood" was attributed to a celestial body, ancients called Phaeton. The ancient people thought it was a comet, a star, because they didn't know the word UFO. Now we know that only UFOs can perform such impossible feats described by witnesses.

Likewise, the Greek writer Apollodorus wrote about the cosmic body that ancients called Phaeton and the great conflagration that followed. "Out-topped all the mountains and his head often burned the stars. Such and so great was Phaeton, when hurling kindled rocks. He made for the very heaven with hissing and shocks, spouting a great jet of fire from his mouth. At mount Haemus he heaved whole mountains. A stream of blood gushed out of the mountains." The Roman writer Pliny the Older described Phaeton as follows: "A terrible comet was seen by the people of Ethiopia and Egypt, to it had a fiery appearance." Also, ancient myths and legends mention a great heat that permeated the earth during a great conflagration. Interestingly, these heat vents are mentioned in many flood traditions. For example, from Persia, there is a legend that mentioned, "The sea boiled and all the shores of the oceans boiled." The cause of the heat was ascribed to the star Tistrya, which also was accompanied by an "incredibly violent hurricane. Let a stream of fire flow toward the earth and filled our world with its devouring heat."

Likewise, several North American Indian traditions refers specifically, to the phenomenon of superheated water. "Great clouds appeared, such a great heat came, that finally the water boiled. People jumped into the streams and lakes to cool themselves and died." Further, on the Pacific coast, the natives have legends that said, "It grew very hot, many animals jumped into the water to save themselves, but the water began to boil." The late great scholar Immanuel Velikovsky wrote that the early traditions of the people of Mexico that were written down in pre-Columbian days' mention what they thought was Venus. "The star that smoked, la strella que humeava, was Sitlae Choloha," which the Spaniards called Venus. Further, the great scientist Alexander Humboldt wrote about this phenomenon.

Now I ask what optical illusion could give Venus the appearance of a star throwing out smoke? The answer is that there are witness accounts of UFOs throwing out smoke. For instance, on October 21, 1978, Frederick Valentich, a twenty-year-old pilot with the Australian Air Training Corps, left Moorabrin Airport in Victoria at 6:20 p.m. to King Island. His flight path would take him across the Bass Strait Triangle, where planes and people have disappeared just like in the Bermuda Triangle.

Valentich was about forty-five minutes into his flight when he contacted the air tower near Melbourne about a speeding aircraft that seemed on a collision course with his single-engine Cessna 182. According to tower controllers, the radar screen was clear. The tower control transcripts go like this:

Valentich: Melbourne, this is Delta Sierra Juliet [Valentich's call sign]. Do you have any known traffic below 5,000 feet?

Tower: Delta Sierra Juliet, no known traffic.

Valentich: I am seeing a large aircraft below 5000. tower: What type of aircraft is it?

Valentich: I cannot affirm. It is four bright lights. It seems to me lake landing lights.

Tower: Delta Sierra Juliet?

Valentich: Melbourne, the aircraft has just passed over me at least a thousand feet above.

Tower: It is a large aircraft. Confirm.

Valentich: Er... unknown due to the speed its traveling. Is there any aircraft from the air force in the vicinity?

Tower: No known aircraft in the vicinity.

Valentich: It's approaching now from due east toward me. It seems to me that he's playing some sort of game. He is flying over me two, three times at speeds I could not identify.

Tower: Confirm you cannot identify the aircraft?

Valentich: It's not an aircraft as it's flying past. It's a long shape. Cannot identify more than it has such speed. It's before me right now. Melbourne, it seems like is stationary right now. What I am doing is orbiting and the thing is just orbiting on top of me. Also, it's got a green light and is sort of metallic like. It's all shiny on the outside. It's just vanished again.

Tower: Confirm the aircraft just vanished.

Valentich: It's now approaching from the southwest. The engine is idling rough. I've got it set at 2324 and the thing is coughing. That strange aircraft is hovering on top of me again. It's hovering and is not an aircraft!

After, a metallic sound was heard, and that was the last transmission of Frederick Valentich. The day was beautiful. The search was done in a fifty-mile radius in sea, air, and land and continued for five days, twenty-four hours a day, and the daylight search continued for three weeks before they called it off. Furthermore, Valentich had four life jackets. Shockingly, no oil slick, no debris were found, nothing. Valentich's father still visits the airport were his son was last time seen, hoping he will come back. He believes that his son still alive.

So we see in this event a clear alien abduction. However, the most important point is that the UFO, it seems, is idling through, just like is mentioned in the Aztec texts, in which they thought the planet Venus throwing out smoke. Undoubtedly, the great biblical Flood was a gravitational-electromagnetic event caused by UFOs. As a matter of fact, whales and other marine animals have been found on the top of mountains. Scientists have reported bones of whales and other cetaceans have been found in various parts of Ventura County in California. Especially along the sides and crests of the Santa Paula Mountains on the summit of this range at an altitude of two thousand feet, scientists have found remains of seals.

Moreover, in the nineteenth century, scientists studying the Lake Titicaca basin in the Bolivian Andes found siluroid, cyprinoid, and other marine fishes in the lake. Further, the great scientist Reginald Daly wrote, "Bones on mountains top: what caused reindeer, bear wolves, and mammoths to leave their bones on the top of the Montagne De Sautenay in Southern France? How can the mountain of bones be explained on the island of Cerigo near Crete. It is a mile in circumference at the base, and from the base to the summit is covered with bones." Interestingly, it seems that the waters of the Mediterranean rose higher and the animals climbed higher and higher, leaving their bones on the famous Mountain Bones. Similarly, the great British Scholar Sir Joseph Prestwick writes, "Throughout Europe are to be found Cath Basins and Ossiferous canyons full of bones and fossils buried in mud."

By the same token, the Himalayas is the highest mountain range in the world, towering more than five miles at the peak of Mount Everest. Interestingly, geologists in the nineteenth century were astonished when they discovered that at the top of the Himalayan peaks, including Everest, were found the skeletal remains of marine life. Indeed, common sense will tell you that these great mountains were once under water. Shockingly, the Himalayas became the highest mountains on earth only during the Pleistocene period, that ended just ten thousand years ago. The Himalayas rebound was due to intense uplift.

Furthermore, as a matter of fact, there is evidence that the continents have been under water. For instance, beneath the suburbs of Paris, geologists have found gypsum deposits containing the remains of more than eight hundred marine species. Geologists, after digging deep down, reach a layer of clay, with reptile bones and fresh water shells. Likewise, the area on which Paris now stands was once under the sea and before dry land as it is today. The sedimentary layers excavated by geologists show a sequence of no less than six changes between sea and land.

In fact, excavations in other European countries reveal the same pattern. Most of the European continent was once under water, similar to North America. For example, the skeletons of two whales were found in a bog in Michigan, and bones of another whale was found north of Lake Ontario. Additionally, in Canada, in an area that is now 440 feet above sea level, more whale bones were found. Also, whale bones were found at Vermont and in a Quebec area at heights of around five hundred feet above sea level. Interestingly, during the colonization of North America, whale bones were found in such numbers in the Alabama soil that farmers made fences out of them.

The finding of whale bones at five hundred feet above sea level raises serious questions above the possible causes of these penetration of the oceans in the continents.

1. Was the whale deposited on high ground by a tidal wave?
2. The ocean floor suddenly erupted; if it did, what was the cause? And how could the interior of the earth create those sudden eruptions? Maybe by the action of volcanoes, carrying the whale with it and stranding the creature in new created land.
3. Or did the whale die and deposit his bones on the ocean floor, which then elevated slowly over millions of years to become dry land? (Although geologists believed that recently formed areas of North America were above water at least 530 million years ago, before whales evolved.)

Moreover, scholars pointed out that are more than eighty thousand legends, myths, traditions, folklore, and religions that mention the Flood. Also, according to traditions and legends, there were three phases in the Deluge or Flood: (1) A pile up of the water in a region of the ocean, (2) a great torrent of water falling, and (3) the final release of the water to the top of the highest mountains, like the Everest. The worldwide traditions agree that the Deluge extinguished all life and a great fire that was burning the world. Furthermore, the myths and legends mention the extraordinary heights reached by the flood waters. For instance, in Genesis is written, "The waters prevailed exceedingly on the earth. And all the high hills that were under the whole heaven were covered. Fifteen cubits upward did the waters prevail and the mountains were covered. The waters stood above the mountains, as tremendous winds caused then to mount up to the heavens, and (God) gathered the waters of the sea as a heap." Similarly, Maori traditions and myths mention "terrible winds that piled up the water's ocean into mountain-high surges." Moreover, ancient Chinese records mention that the waters, "in their vast extent over-topped the great heights, threatening the heavens with their floods." By the same token, a Midrashic legend mentions, "The waters were piled up to a height of 1600 miles, and they could be seen by all the nations of the earth!" Further, the Lappish legends describe the destruction of a wicked world by a very angry deity named Jubmel who said, "I shall reverse the world; I shall cause the sea to gather together itself up into a huge towering wall, which I shall hurl upon your wicked earth children and thus destroy them and all life!" Most legends mention that waters piled up to heights so great that they "overtopped" heavens themselves, water gathered into a huge towering wall.

The legends and myths about the flood are very consistent. The myths and legends mention of a sudden rise of the ocean waters, and it does not fit with any uniformitarian theory of the earth. The myths agree that an immense wall of water gathered together at a particular locality in the ocean and was held there by an immense gravitational force.

Interestingly, in the Bible, in Psalms, is mentioned that the watery wall "stood fast" or it was immobile. This phenomenon is not explainable in any known normal workings of nature. This phenomenon is explainable only if a second gravitational field were acting in opposition to the terrestrial gravity and magnetic fields. Such a force would cause rivers to flow upward and which heaped up the ocean waters and held them stationary for a lapse of time. The opposing gravitational field of UFOs can cause these effects.

Further, the Babylonian Talmud mentions that seven days before the Deluge, the holy one changed the primeval order. Many scientists believe that the axial stability and rotational speed, with water can be changed running uphill rather than downhill. As a matter of fact, as far back as 1936, scientists considered the effects of a "hypothetical celestial influence which suddenly decreased earth's rotational speed and caused a rapid change in the shape of the hydrosphere, inferred that sea levels would be greatly depressed in low equatorial latitudes and drastically raised in polar latitudes." In fact, this is precisely what the legends of the world mention. Obviously, it seems to imply that the piling of the ocean's water has occurred at or around the poles or at the pole only, the northern one. The legends refer to the Deluge waters as having flowed strongly northward; it seems that these waters were attracted to the polar hemispheres. Also, the weight of the accumulated water at the pole may have been the cause of an axial tilt, at the near proximity of a powerful external influence of a UFO. Indeed, we can say that this great event in earth history was not caused by a cosmic event or a prolonged meteorological event but was caused by a UFO. This event was a gravitational and electromagnetic effect caused by a UFO or UFOs, which as consequence caused floods, excessive rainfall, evaporation, and supersonic winds. Certainly, the inhabitants of high latitudes experienced the most massive rising of the deluge waters, as they became piled up into "enormous walls of water mountains submerging all land."

Interestingly, some myths mention arks and other floating devices as the only way to escape the rising waters. Furthermore, the Bible and other myths and legends mention forty days of rain that, with the flood, contributed to the submergence of the land. By the same token, the Apocalypse of Thomas mentions enormous blocks of hail (ice) showering the earth, just before the Deluge. In fact, the Persian myth of Zend-Avesta mentions enormous drops of water, and sometimes they were boiling. The myth mentions that the drops of water were the size of a man's head. The American Indians also mention raindrops of phenomenal size. Obviously, the water rained in drops so large the origin and formation cannot have been due to even the most dust-polluted atmosphere or greenhouse gases but a creation of the UFO anti-gravitational and electromagnetic fields. Moreover, the Genesis, the Tepanecas of pre-Columbian Mexico, the American Indian tribes and hundreds of myths and legends mention a "rain that has ever fallen before or since" because it has never been seen. The rain was described as a "seven water torrents, an overflowing rain, great hailstones descended out of the clouds, as if poured out of jars, like bulls and man's head." According to Genesis, the rain fell torrentially and uninterruptedly for forty days and nights. The rain swamped all the lands and mountains; it was not an ordinary rain, because even many early writers saw the difference between a seasonal rain and an exceptional cataclysmic deluge. Unfortunately, the Bible and other ancient myths and legends made emphasis on the fury of God or "divine retribution." Although, it is more likely that the real cause of the Deluge was the Intelligence behind the UFO phenomenon. Moreover, the Norse myths, the Eddas, mention that the heavens have been shattered falling in the form of "hail and huge blocks of ice." Likewise, the Maori tradition mentions that "the celestial waters pour down and flood the earth."

Interestingly, the Jewish myths mention a date when the cataclysm occurred. "On the seventeenth day of the month of Cheshvan, the male waters fell from the heavens while the female waters welled forth from the depths. They united and waxed strong and overwhelmed the earth and all that was upon it. Cheshvan was the second month of the old Jewish year.

Similarly, Genesis mentions that the event occurred on the seventeenth day of the second month. Undoubtedly, the American Indian legends mention how the advance of the waters was watched for hours and days before the waters overwhelmed the land. For instance, the Choctaw Indian tribe of Oklahoma mentions in their legends that after the primeval earth had been plunged into darkness for a long time, a bright light (UFO) appeared in the north, which occasioned great joy. At the same time, a mountain-high wall of water advanced and destroyed the land. Also, the Navajo Indians of Arizonain one of their legends mention how, long ago, there appeared in the east, and subsequently in the south, north, and west, a phenomenon which from a distance resembled a high, steep wall of rock. But actually it was water slowly advancing toward them. Additionally, another Navajo legend mentions interesting details about how their ancestors were surprised one day to see animals of every type running from east to west, four days later, they saw a bright light (a UFO) in the east." So they sent few people to investigate and returned with the news that it was a vast flood of water proceeding in their direction.

The next morning, a water mountain had filled the horizon except the west, which "advanced like a chain of high mountains." Undoubtedly, all these traditions, legends, and myths agree that the flood waters were amazingly high and advancing as a watery wall. As a result of the UFOs holding the water with anti-gravitational and electromagnetic fields, the wall of water, toward the equator, slowly and rapidly gaining speed of more than 800 mph. Interestingly, even the Koran mentions that the Deluge was simply a vast wave. Genesis mentions the Deluge: "And the Lord said unto Noah, come thou and all thy house into the ark . . . For yet seven days, and I will cause it to rain upon the earth forty days and forty nights, and every living substance that I have made will I destroy from of the face of the earth . . . and it came to pass, after seven days, that the water of the Flood was upon the earth (Genesis 7:1, 4, 10). Similarly, the Greeks mention the Flood and their hero they called Deucalion. In Australia, India, Polynesia, South

America, Europe, and Africa, legends and myths about the Flood have been handed down since the beginning of history. As a matter of fact, all the Flood legends, myths, and traditions reflect the same worldwide catastrophe.

Interestingly, at the Sumerian city of Ur, archaeologists have found the remains of a gigantic flood, which had deposited a layer of clay almost ten feet thick, containing traces of human habitation. As a result, archaeologists came with an approximate time when the great Deluge took place. It took place 4,000 BC. Furthermore, how did marine fossils get up on top of mountains? Fossils of corals have been found on mountaintops, four thousand feet above sea level. For instance, in the island of Timor, fossil sea shells and sea lilies are found thousands of feet above sea level. Interestingly, a whale skeleton was once found a mile high on the California coast range Also, on the top of Mount Sanhorn in the arctic coast, a skeleton of a whale was found. My question is, did the worldwide myth of the flood mention in the Bible and other eighty thousand legends, myths, traditions, and folklore around the world, is responsible for these fossil findings? Did a universal flood leave them there as the waters receded? If these whales died in the water while the mountains were still submerged in a geosyncline, how could the fossil bones survive the upward passage through the waves? How could the mountain itself survive as it rose past the waterline? Would the natural elements like the wind, the tides, and the waves grind the bones to powder and the rock to sand? In my opinion, the most obvious answer to these puzzle is that the ocean rose to the top of mountains in the recent past.

Moreover, at British Columbia, the trilobites are on top at the very highest part of the rocky mountain. Just before the railway goes through the tunnel at Kicking Horse Pass enormous beds of fossil trilobites are found. In fact, carloads of them have been removed from Canadian Pacific freight cars and sent to the many geological laboratories and museums around the country. Shockingly, these rocks are labeled Cambrian and are 440,000 years old because trilobites are primitive invertebrate fossils of the Cambrian. Moreover, with the trilobites were found sponges, brachiopods, worms, jellyfish, and shrimps like crustaceans.

Furthermore, more than 70 genera and 130 species have been found high above the timberline on the planks of Mt. Wapta in the Rocky Mountains. Certainly, the problem is why has Mt. Wapta not been eroded to sea level in 440,000 years? So geologists answer that often the land was submerged and marine sediments were deposited and elevated to form local mountains, which were again worn down. Land was uplifted during periods of mountain building and destruction sixty million years ago. However, my question is, why were not all these soft bodies destroyed during the uplift of the mountains? How could this isolated bit of sea bottom now be found above the timberline in the Rocky Mountains?

Furthermore, a great scientist, Dr. L. Stokes writes, "There is a profusion of soft-bodied animals, so delicate and unsubstantial that their preservation seems scarcely possible." For example, scientists have found worms, jellyfish, shrimp, and other crustaceans, none of which have hard shells or skeletons and are amazingly preserved. As a matter of fact, some organisms are so excellently preserved that they show internal organs like films of carbon! Conversely, this whole area is part of the Lewish Overthrust, which according to geologists is supposed to have shoved these 440,000.000 years old! Cambrian shales, with the sponges and jellyfish up over the cretaceous limestone, are only 100,000.000 years old!

To me, it is a puzzle to explain how this thin bed only a few feet thick, with its sponges, jellyfish, and 130 species that somehow got up on the top of Mount Wapta. However, the main problem is how to account for the trilobites at the very top when classical geological theory insists that the trilobites should be at the very bottom! Obviously, we can't explain these facts of nature in any normal form we know because it is beyond our experience. So there are only two ways to answer this: either mystery or nature worked in a different form millions of years ago, or there is something supernatural in these events and the only possible cause is the intelligence behind the UFO phenomenon.

We don't know of any force in nature that can exchange bedrocks of different geological ages. I offered an explanation for these phenomena in chapter 3. Moreover, when the great Deluge took place accompanied by earthquakes around the world, from the Pacific across South America to the shores of Africa, traditions, legends, myths, and religions mention that in the night skies a giant comet that shone (or aerolite or wandering star, or planet—nobody knows what kind of object—or a UFO?) brought destruction on Planet Earth. According to legends, it was a luminous object that flashed across the sky, although there is no record in historical times of celestial bodies falling from the sky.

According to archives of the ancient temples of Thebes and Egypt, the Deluge and the disaster that preceded and ended were witnessed by the ancestors of the human race. The legends mention "the greatest earthquake that has ever shaken and convulsed earth." Night and day the earth shook, and a rain of fire came from the heavens that set ablaze the forests. Interestingly, the Indians of South America heard terrifying noises coming from heaven; it seemed as if giants in the sky were bombarding the earth with a deluge of rocks. The light of the sun appeared to have gone out because for days darkness enveloped the earth. After, an immense cloud of reddish powder filled the air after a rain of fine cinders, which covered the trees. According to the South American Indian legends, four tremendous explosions shook the ground and four enormous white-hot globes fell from the sky into the forest. Next, the river began boiling in a hissing steam and rising to the atmosphere, increasing the heat of the forest.

Interestingly, legends all over the planet mention these same disasters. Similarly, in the Sahara Desert, the inhabitants mention traditions of a great disaster that created the Sahara Desert and destroyed ancient civilizations. For instance, they have stories of rain that lasted for forty days and forty nights. Moreover, the South American Indian legends mention "great and terrible rumblings both and below ground." Also, "The sun and stars burned red, blue and yellow and wild beasts mingled fearlessly with men." The legends mention, "The Indians heard a roar and saw darkness ascending from the earth to the sky, while thunder rolled terrifying the people and the rising waters submerged the land."

Interestingly, in Egypt, in the great hall of the temple of Rameses at Karnak (Thebes) by the Nile, a legend is written about the feast of renewal, celebration, and mourning for the loss of a drowned continent in the western ocean. Similarly, in Mexico, the Aztecs celebrated a feast of the renewal of the human race. Every four years, the Aztecs and Central American Indians added another eight days in memory of the three occasions that the world perished. The similarities between the two legends are striking even though Egypt is more than 6,400 miles away. The Aztecs recorded three great catastrophes, of which the third was the greatest. As a result, each catastrophe was followed by an era of ruin and destruction of the human race. Likewise, the Aztec Codex Chimalpopeca mentions a "rain of fire, which followed the sun of rain, all that existed burned and there fell a rain of rocks and sandstone. The sky drew near the water and in a single day, all was lost. The mountains themselves sank under water, but the water remained calm for 52 springs."

According to the ancient writings, "Men climbed trees, ran everywhere, in their terror, crowding and pushing together, embarked in ships, hid themselves in caves, got on mountain tops." Certainly, the few survivors were so far away that they thought there were the only human beings in the planet. Interestingly, every four years, a celebration was held in Mexico and Central America. In the celebration, the royalty and the peasants prostrated themselves to the gods to prevent more catastrophes. Moreover, it is mentioned in the legends that periods of darkness covered the earth that lasted up to twenty-five years." Then, men "were lost and carried away in a rain of fire, all was consumed and all the Lords perished." In fact, evidence of this rain of fire is found near Mexico City, in the great lava bed, Pedregal de San Agustin; under the bed, archaeologists have found ancient houses and pottery.

The Quiche legends of the Indians of Guatemala in the book of Popol Vuh mention how the people "were lost in the terrible flood, followed by thick rain of bitumen and resin, when ran, here and there and madness." Conversely, the people tried to climb the roofs of their houses, but they crumbled with the extra weight. Also, they tried to hide in caves and grottoes, but they were shut in from the exterior. Moreover, the legends mention that the earth darkened and it rained day and night; the survivors went insane after so many frightening natural disasters. Similarly, the natives of North America, British Columbia, and Alaska mention in their legends that "the forefathers of the natives, took to their canoes, which were ultimately stranded on the sides of the mountains. Bears and wolves swam off to board the boats and were driven back by clubs and spears. Our folk landed right on the mountain-top and set to erect walls of great height to keep the rising water. Here they docked their boats and watched floating past, great trees torn up by the roots, and monstrous devil-fish and other strange and terrible creatures of the land and sea."

By the same token, in the region of the upper Maranon or Amazon, the native Indians mention a legend in which they were terrified by a "deluge of hot and steamy water, which poured down on the earth, burning and scalding it all up and destroying the forest. On earth all was dark as night for many moons. The sun was completely hidden for days." Also, the Indians of Brazil have tradition of a great flood that covered the whole earth inhabited by their forefathers. Also, the eastern coastal range of Brazil remained above water. Likewise, the Ipurina Indians of the River Purus in the Upper Amazon mention a deluge of hot water that fried the fish! The scalding liquid poured down on earth burning everything, even the forests, and the earth was in darkness.

Furthermore, in the same region of the Amazons, the Parrarys, Abederys, and Katausnys Indians mention a legend of catastrophe. "Once upon a time, folk heard a great rumbling above and below ground. The Sun and Moon turned red, blue and yellow and the wild beasts mingled fearlessly with men. A month passed and our forefathers heard a roar and saw darkness ascending from the earth to the sky, accompanied by thunder and heavy rain which blotted out the earth and made day into night."

I describe in chapter 4 the object that exploded in 1908 in the Russian Siberian, which produced the same effects described by the aborigines in their legend in Brazil. So the South American legends and myths agree that the waters rose very high, till the earth sunk beneath and only the branches of the highest trees stood out above the flood. The people died of cold and hunger, and only Uassu and his wife were saved, when they came down from the heights. Furthermore, after the Deluge, another catastrophe took place on the Andes that raised them to the present heights. As a result, the city of Tiahuanaco rose to twelve thousand feet up. Similarly, in Bogota, Colombia, there are what are called the giant fields. This place is filled with fossilized and petrified bones of mastodons, mammoths, and Pleistocene animals. Interestingly, the Andean plains rose more than two miles high!

Is Phaeton an asteroid or a UFO? As I have shown, ancient civilizations mention in their myths and legends a celestial object that they called Phaeton, although everything now we know about this celestial object points to what the ancients describe are UFOs. Moreover, the legends and myths agree that Phaeton caused a major destruction of the world. For instance, the natives of the Samoan islands mention, "The sea arose and in a stupendous catastrophe of nature, the land sank into the sea. The new earth [The Samoan Islands] arose out of the womb of the last earth."

Similarly, a Tahitian legend states, "In ancient times Taaroa, the principal god, the creator of the world, being angry with men on account of their disobedience to his will, overturned the world into the sea, when the earth sank into the water, excepting a few aurus [projecting points] which remained above its surface constituting the present clusters of islands." As a matter of fact, the last two traditions suggest that these Pacific islands were once part of a large now-vanished continent.

Interestingly, a pre-Columbian Maya manuscript known as the Troana Codex mentions its destruction. "There occurred terrible earthquakes, which continuously until the 13th. The country of the hills of mud, the land of Mu, was sacrificed, being twice upheaved it suddenly disappeared during the night. The basin being continuously shaken by the volcanic forces." Also, the legends mention that the land sank and rose in several places and "ten countries unable to stand the force of the convulsion, they sank with their 64'000.000 of inhabitants 8,060 years before the writing of this book."

After all, true or false, there are other legends that mention this disaster. For instance, the Hawaiian tradition mentions that it was a great continent. "Stretching from Hawaii, including Samoa, Lalakoa and reaching as far as New Zealand. Also, taking in Fiji and there were some low lands in between. These higher lands, all this was called by one name Ka-Houpo-o-Kane. The solar plexus of kane, and was also called Moana-Nui Kai-oo or the great engulfing ocean." The aborigines of Mexico, called the Mixtecs in their myths, mention of a now-vanished "land to the east of the American coast. In a single day all was lost, even the mountains sank into the water. Subsequently there came a great Deluge in which many of the sons and daughters of the gods perished." From the Indians of Peru, a legend mentions, "in those days of the Jaguar-faced gods called Huaca, the Andes were split apart and the callejon was formed, when the sky made war on the earth." Moreover, from the Indians of Brazil, a legend mentions, "The lightning flashed and thunders roared terrible and all were afraid, then the heavens burst and the fragments fell down and killed everything and everybody. Heaven and earth change places, nothing that had life was left upon the earth."

Additionally, the Kato Indians of California mention a legend: "Every day it rained. All people slept and the sky fell. The land was hot; the waters of the oceans came together. Animals of all kind drowned, where the water went there no trees. Human beings and animals alike had been washed away. It was very dark." By the same token, the Bible and many myths and legends worldwide mention a completely darkness that enveloped the earth. The Aztecs mention this phenomenon.

"The third sun is called Quia-Tonatiuh, sun of rain, because there fell a rain of fire, all which existed burned and there fell a rain of gravel. Now this was in the year Ce Tecpalt. Now in this day, in which men were lost and destroyed in a rain of fire, they were transformed into goslings. The sun itself was on fire and everything together with the houses was consumed. A tremendous hurricane that carried away trees, mounds, houses, men and beasts." Many people escaped the great hurricane hiding in caves. All this time they were in darkness, without seeing the light of the sun, nor the moon. "After, the destruction of the fourth sun, the world plunged into darkness during the space of twenty-five years. Amid the profound obscurity, ten years before the appearance of the fifth sun, mankind was regenerated."

Similarly, in British Columbia, Canada, the Akawais Indians remember in their myths a period of prolonged darkness and the intense cold it caused at the time of an immense world flood. The great scholar De Bourbourg wrote, "A vast night reigned over all the American land, of which tradition speaks unanimously. In a sense the sun no longer existed, for this ruined world which was lighted up at intervals only by frightful conflagrations, revealing the full horror of their situation to the small number of human beings, that had escaped from these calamities."

Similarly, the Samoan traditions mention, "Low clouds appear to envelop the world and plunged into darkness. It was during this period that the islands of Samoatonga, Rotuma, Wallis and Fotuna were upheaved from the bed of the ocean." During these events the Samoans believed that the heavens fell down and were so low that people could not stand upright without touching them. Moreover, with the Deluge, there are many accounts of hail and fire in many myths and legends. One legend describes, "And every island fled away and the mountains were not found. And there fell upon men a great hail of heaven, every stone, and men blasphemed god, because of the plague of hail. For the plague thereof was exceeding great."

Antigravity and electromagnetism in UFOs can cause all these effects described in legends and myths. UFOs create the energy field of antigravity by creating an artificial space and time curvature; that is the reason we find fossils of different eras. So the UFO creates an artificial mass with a gravity as strong as the earth or stronger. Also, this is the technology used to trigger the Deluge and hold the oceans in a wall of water. Hydromagnetics plasma in physics is the study of the reaction of plasma fluids to high magnetic fields. Interestingly, ionization and magnetism combined produced a hydromagnetic effect, in this case the biblical Deluge. The UFO is a rapidly pulsating magnetic field; the charged particles moving across the field would experience a deflecting force that would cause the oceans to pile up in a great wall and invade the continents. Also, the magnetic effect would reverse the magnetic fields and redirect the magma so that it would cause continents to sink or to rise.

The late great scholar Immanuel Velikovsky stated that a close approach to the Earth by a celestial body, which he identified as the planet Venus or Mars, would have caused the last Pleistocene mass extinction.

Moreover, according to myths and legends, a celestial body took the form or shape of lions, jackals, dogs, pigs, and fish. Although we know that comets or asteroids can't take the form of animals, we know that UFOs can take any form or size. For instance, the astronomer Guy Murchie in the sixteenth century in his book Music of the Spheres wrote a description of the great comet of 1528. "It looked so horrible and produced such a great terror in the common people, that some died of fear and others fell sick. It appeared to be of excessive length and the color of blood. At its summit, from its nucleus rose the figure of a bent arm, holding in its hand a great scimitar as if about to spike. On both sides of the rays of this comet, there appeared a great of axes, knives and blood drenched swords. Among which many hideous slowly-shifting faces with beards and bristling hair." Obviously, we know that stars, comets, planets, asteroids and meteorites do not take or change form. We know however, that UFOs can take any shape or form.

Interestingly, UFOs can change the natural environment. For example, at the so-called Miracle at Fatima in 1917, approximately seventy thousand people witnessed the event. Among them were clerics, journalists, policemen, military personnel, and teachers. Unfortunately, people though that it was a god appearing in a supernatural form, but know we know that it was a UFO.

During one of the apparitions, thunder and lightning preceded the materialization of the Madonna. The phenomenon was centered on three children—Lucia, Jacinta, and Francisco. Lucy reported the apparition of an angel: "He was holding a chalice in his left hand, with the host suspended above it. Some drops of blood fell into the chalice. Leaving the chalice suspended in the air." So we see in this event that the intelligence behind the UFO phenomenon can nullifies gravity.

In another supernatural event, the angel appeared to the three children. A great blast of wind preceded his appearance. Lucy wrote, "We were playing only for a few moments when a strong wind began to shake the trees. We looked up startled to see what was happening. For the day was usually calm, then we saw coming towards us above the olive trees, the figure I have already spoken about. As it drew closer, we were able to distinguish its traits. It was a young man, whiter than snow." Furthermore, while this was going on, "the crowd was so silent that you could have heard a pin drop." Two witnesses and some others that were close to the children heard "an unintelligible murmur, then I began to hear a sound, a little buzzing, rather like a mosquito in an empty bottle, but I couldn't hear any words." This mysterious murmur was heard by only a few witnesses. However, two other unusual phenomena were noticed by much greater number. "The luminosity of the sky noticeably decreased, as during an eclipse!" The whole time, the ecstasy of the children lasted. At the same time, "the temperature which was very hot, went down noticeably, and the tint of the light was modified. The atmosphere became yellow as gold." There was a sudden stopping of the rain, although it had been raining all morning. Shockingly, during the apparition of the Madonna, the rain totally stopped and suddenly the sky cleared.

The sudden change in the weather surprised all the witnesses. Another witness said, "It was a rainy day, but a few minutes before the Madonna appeared, the rain stopped." Another witness reported, "At this moment I got out of the car, and as extended my hand to my wife to help her step out suddenly all the clouds disappeared without the slightest breeze and the sun was shining in a clear sky." Moreover, a

ESSAY ON THE THEORY OF THE EARTH: ELECTROMAGNETISM IN UFOS AND THE ORIGIN OF MASS EXTINCTIONS AND THE ICE AGES

local newspaper reporter stated, "An unbelievable spectacle for anyone who did not witness it; from the road. One could see the immense multitude turn towards the sun, which appeared free from clouds and in its Zenith it resembles a dull disc. And it is possible to look at it without the least discomfort. It might have been an eclipse which was taking place." Similarly, Dr. Almeida Garret, who observed the phenomenon, said, "I veered to the magnet which seemed to be drawing all eyes and saw it as a disc with a clean cut rim, luminous and shining." Another witness said that the object was "a dull silver disc"; also "it was a clearer, richer, brighter color, having something of the lustre of a pearl." Another description of the object was, "It was a living body. It was not spheric like the moon, it looked like a glazed wheel, made of mother-of-pearl."

As a matter of fact, witnesses said, "It could not be confused either with the sun through fog [for there was no fog at the time] because it was not opaque, diffused or veiled. It gave light and heat and appeared clear-cut with a well-defined rim. The sky was mottled with light cirrus clouds, with the blue coming through, here and there. But sometimes the stood out in patches of clear the sky. The clouds passed from west to east and did not obscure the light of the sun, giving the impression of passing behind it, though sometimes these flecks of white took on tones of pink or diaphanous blue as they passed before the sun." Another witness reported, "It was a remarkable fact, that one could fix one's eyes on this brazier of light and heat without any pain in the eye or binding of the retina."

Further, the witnesses looked up the sky in fear, thinking the world was coming to the end. They thought that the sun was performing some kind of a wild dance. The witnesses reported, "Suddenly, the sun began trembling, shaking with sudden movements finally tuning over upon itself with a dizzying quickness, spraying out rays of light of all the colors of the rainbow. The sun trembled, the sun made sudden incredible movements outside all cosmic laws." The sun "danced," according to thousands of witnesses. Likewise, another witness declared, "It shook [the sun] and trembled, it seemed like a wheel of fire." Many witnesses give the same account. Also they said, "It spun like a fire wheel, taking on all the colors of the rainbow" and "It looked like a ball of snow, revolving upon itself." Another witness said, "The sun's disc did not remain immobile. This was not the sparkling of a heavenly body, for it spun round upon itself in a mad whirl."

By the same token, Father Pereira da Silva wrote about the miracle. "The sun appeared with its circumference well-defined. It came down as if to the height of the clouds and began to whirl giddily upon itself like a captive ball of fire, with some interruptions this lasted about 8 minutes." Another witness reported, "At a certain moment the sun seemed to stop and then began to move and dance. However, the sun stops, only to begin its strange dance all over again, after a brief interruption, whirling upon itself, giving the impression of approaching or receding." In fact, the dance of the sun that seventy thousand people witnessed was not the sun but a UFO. The sun cannot do these types of acrobatics that were repeated three times. Also, changes in the atmosphere were reported by witnesses. "I saw everything an amethyst color, objects around me, the sky and the atmosphere were of the same color. An oak tree nearby threw a shadow of this color on the ground." Similarly, another witness reported, "The sun took on, all colors of the rainbow, everything took on the same colors, our faces, our clothes, the earth itself. A light whose color varies from one moment to the next, is reflected on persons and things." Additionally, witnesses reported the falling of the sun: "Then suddenly, one heard a clamor, a cry of anguish breaking from all the people. The sun whirling wildly, seemed to loosen itself from the firmament and advance threateningly upon the earth as if to crush us with its huge and fiery weight." Another witness said, "The sensation during those moments was terrible, the sun began to move and dance until it seemed that it was being detached from the sky and was falling on us. It was a terrible moment. It seemed like a wheel of fire was going to fall on the people." Another witness described, "It suddenly seemed to come down in a zig-zag menacing the earth." Similarly, another witness said, "The sun at its zenith whirled upon itself, it detached itself in descending towards the right and the left;

having almost arrived at the horizon line, it went back up to the zenith on the left, tracing a sort of winding ellipse as it went."

Many people fainted from the terror of seeing the sun falling. Shockingly, the dance of the sun (a UFO) was seen twenty-five miles away from the area of Cova da Iria, where the miracle took place. Also, Father Ignacio Lourenco reported, "I feel incapable of describing what I saw and felt. I looked fixedly at the sun, which seemed pale and did not hurt the eyes. Looking like a ball of snow revolving upon itself, it suddenly seemed to come down in zigzag, menacing the earth. Terrified, I ran and hid myself among the people, who were weeping and expecting the end of the world at any moment. During those long moments of the soar prodigy, objects around us turned all the colors of the rainbow." Moreover, the witness proceeded, "Everybody's clothes got dry! A last astonishing fact, all those people, who were for the most part soaked to the bone, noticed with joy and stupefaction that they were dry!"

The academician Marquez Da Cruz wrote, "This enormous multitude was drenched, for it had rained unceasingly since down. But though this may appear incredibly, after the great miracle everybody felt comfortable and found his garments quite dry, a subject of general wonder. The truth of this fact has been guaranteed with the greatest sincerity by dozens and dozens of persons of absolute trustworthiness, whom I know intimately from childhood." Likewise, another witness reported, "The sun seemed to be loosened from the sky and to be approaching the earth, strongly radiating heat as though, we had entered an overheated steam room." Obviously, what seventy thousand people thought was the sun dancing was a UFO. The sun cannot perform acrobatics of this kind. The sun doesn't go around the solar system, but the planets go around the sun.

Interestingly, in 1505, a supernatural event took place in Poland. There was a missionary of the Franciscan order named Ladislaw during the time of the war between Poland and the Tartars and the Turks. The missionary called upon the people to pray and have faith in God. Then, while they prayed, the invading army encamped between the Pruth and the Dniester rivers. Suddenly, the waters rose in a flood, inundating a large area, and this was followed by a severe frost and then by a blinding snowstorm. As a result, most of the enemies' troops perished, and Poland was saved. Obviously, the Deluge, the Ice Ages, and mass extinctions are completely contrary to the normal workings of nature. Undoubtedly, there is no terrestrial mechanism capable of generating on its own a cataclysm as worldwide as the Deluge. My conclusion is that the only external force capable of creating such cataclysms is the intelligence behind the UFO phenomenon.

CHAPTER 3

The Origin of Boulders, Nature's Puzzles, and the UFO Phenomenon

> If man was to think beyond what the senses had directly given him, he must first throw some wild guess-work into the air, and then, by comparing it bit by bit with nature, improve and shape it into a truth.
> —William Smith, Thorndale, 1859

The Origin of Boulders

One of the great mysteries of geology is how a 10.000-ton rock had been transported and left hanging in cliffs and mountain tops around the world. In fact, there are few theories about the possible transportation of these giant boulders: (1) The theory that boulders were transported by a great flood, (2) the theory that ice transported the boulders, and (3) the theory that glacier waves transported the boulders.

Let's examine the theory that a flash flood may have transported the boulders. In my opinion, it is impossible for a boulder ten or twenty thousand tons to float, because even a two-pound rock sinks in the water. The boulder and the rock have no buoyancy; for water to be able to transport a 20,000-ton boulder, it has to be at supersonic speed, almost instantaneous. We know that the water has no physical properties capable of carrying solid objects of the weight of a boulder, nor even a 100-pound rock! For instance, in Conway, New Hampshire, there is a 10,000-ton boulder.

Moreover, sometimes boulders are found from a few miles to thousands of miles from their original source. Certainly, a flash flood going at supersonic speeds will destroy the boulder by impact. Similarly, the boulder found in Warren County, Ohio, has an area of 20,000 square feet and is a 13,500-ton slab of limestone. Shockingly, it averages only 5 feet in thickness. My question is, how did the flood or the glaciers transport this boulder without breaking it to pieces?

Furthermore, in a few cases, material from a single source has been found to litter the landscape for hundreds of miles, forming either linear or fan-shaped boulder trains. One famous boulder train originated from an island off the southwestern coast of Scotland. It fans out over an angle of 150 degrees and extends over a distance of up to 300 miles. Some of its stone is being deposited as high as 250 feet above their point of origin. Many geologists scratch their heads trying to explain this phenomenon. They think they had been laid down by a slowly meandering glacier that changed its direction of advance. Other scientists postulate the theory that these boulders were dispersed by the onrushing of turbulent waters of a glacial wave.

The glaciers cannot carry ten—or twenty-thousand-ton boulders because there is a good possibility that they will be pulverized by the weight. Similar are the erratic British boulders of porphyry, limestone, and sandstone. According to geologists, because the Weichselian ice sheet alleged to have transported them, it is commonly thought as having advanced from the direction of Norway to the northeast although those boulders can only have traveled from the northwest against the supposed direction of the Weichselian ice flow.

Similar cases occurred on both sides of the Atlantic. For instance, in Labrador, erratic boulders have been rammed into hillsides with violence. Moreover, erratic boulders have been found in the Sahara Desert, on the Mongolian plains, and in the country of Uruguay. We must seriously question the Ice Age theory because in numerous instances, the places where the boulders are found are far outside the traceable limits of the supposedly bygone ice sheets.

Likewise, the Ayers Rock is the world's most spectacular erratic boulder and the largest. The Ayers Rock is situated in the northern territory of Australia. The boulder is dark red, especially red at sunset. It is 2¼ miles long by 1¼ miles wide and with a height of 1,143 feet above the ground. It is called the Red Heart of Central Australia because it is composed of red rock, shaped like a heart, and is located near the center of Australia. Interestingly, there are no nearby mountains from which it could have been carved out. Indeed, there it is, in the middle of the desert, with the sage brush and the cactus, a huge boulder for which modern geology can't offer a good explanation of its origin. No iceberg or flood can transport these boulders. The Ayers Rock must have been rafted from somewhere, as neither the conifers found with the rock and certainly not the icebergs are native to Australia. By the same token, the suggestion that ordinary icebergs transported this giant boulder is easy to refute, because if you look at a map of Australia and note the dotted line marked "northern limit of drifting ice," it means that no ordinary-sized iceberg ever gets past the limit, and if it did, it would be cast upon the shore and melt. Certainly, never it will get into Central Australia.

Some geologists believed that there was an extraordinary iceberg that transported and dropped the Ayers Rock in Central Australia. As a matter of fact, icebergs can't transport or drop boulders because they are not steady vehicles. Furthermore, in many places of the world, large boulders are found in a position that proves that a great force must have lifted them up and carried them long distances before depositing them where they are found today. Interestingly, many boulders are of entirely different mineral composition than the soil where they are found. Shockingly, some boulders weigh as much as 10,000 tons. In the nineteenth century, scientists believed that enormous tides had swept over the continents and carried with them masses of stone. The transfer of the rocks was explained by the tides, but what could cause those billows to rise high over the continents?

Similarly, other scientists offer another explanation. A series of gigantic waves was mysteriously propagated. These waves were supposed to have precipitated themselves upon the land and then swept madly on over mountains and valleys alike, carrying along with them the boulders. The stones and boulders on the hilltops are explained as the result of the waves. In fact, there is no evidence that these types of spasmodic waves exist and have ever been experienced by humankind.

The great Louis Agassiz, the discoverer of the Ice Ages, believed that ice sheets had carried the boulders through great distances. For instance, the Madison Boulder near Conway, New Hampshire, measures 90 × 40 × 38 feet and weighs almost 10,000 tons. It is composed of granite quite unlike the bedrock beneath the boulder. Louis Agassiz's theory is that the boulders have been carried by great continental ice sheets. The glaciers of the Alps moved the boulders down the slope, not upward; that is why there is no explanation for the boulders found on hilltops. Moreover, large boulders would sink into the ice sheet.

As a matter of fact, geologists can establish by the extent of denudation of the rocks under the erratic boulders that the boulders were deposited at their places during recent human history. For instance, in Wales and Yorkshire, where the soil effects of the boulders were evaluated, the consensus by some geologists is that no more than six thousand years has elapsed since the boulders were left in their position. Obviously, the fact that accumulations of stones were transferred from the equator toward the higher latitudes is an enigmatic problem to solve for the Ice Age theory.

In the northern hemisphere, in India, the moraines, boulders were carried from the equator, not only toward higher latitudes, but also toward the Himalayas mountains in the southern hemisphere, from the equatorial regions of Africa toward the higher latitudes across the prairies, deserts, and forests of the African continent. In the early nineteenth century, it was baffling to science to rationalize the mystery of the erratic boulders. Scattered over northern Europe and North America were thousands of boulders of different sizes, which were clearly out of place. In fact, the nearest bedrock like them was often hundreds of miles away. Something must have moved them all these long distances, but no one seemed to know what had the power to move boulders weighing thousands of tons. Interestingly, Irish peasants believed that the Viking invaders had brought them with their ships. On the other hand, scientists didn't offer a better explanation. Many in the scientific establishment simply gave it up. After all, wind and water certainly could not have transported boulders that often weighed ten thousand tons!

So utterly baffled were they, in fact, that Charles Darwin mentioned a story about a man named Mr. Cotton "who knew a good deal about Rocks" of whom he remarks, "He solemnly assured me that the world will come to an end before anyone would be able to explain how this stone came where it lay." Referring to one of these huge erratic. Many geologists of that day, in this as in other matters, turned for explanation to Noah's Flood. They assumed that somehow swift currents had set in, lifting and carrying these massive boulders as if they were corks. Just in itself that seemed hard to believe but became impossible when the positions of some erratic were considered. Because shockingly, in the valleys of the Alps and Jura mountains, such boulders were often poised high on the steep slopes and in such precarious balance that any swift rush of floodwaters could not have left them there but would have tumbled them, rolling down the valley. Shockingly, there were thousands of boulders left in unstable places.

Nevertheless, another theory to explain the origin of erratic boulders was postulated by the German geologist Leopold, Baron von Buch. His theory was that nearby mountain ranges had suddenly sprung up, catapulting these rocks to great distances. But how could erratic boulders, falling thus in violent bombardment, have possibly alignedso gently as not roll down into the valleys? Moreover, in 1830, the great English geologist Sir Charles Lyell proposed a theory—that the regions where the erratic boulders are had at one time been submerged beneath the sea. Also, according to his theory, icebergs floating down from glaciers carried rocks and rubbish, including boulders. He wrote, "When the icebergs melted, their load of rocks was dropped, in these strange positions. Presumably sinking gently through the buoyant waters to their present places on what was then sea bottom." For a time, scientists and scholars found this theory not completely satisfactory but at least tolerant. Interestingly, now it is accepted that in few cases where the land has been submerged below sea level, icebergs could have carried erratic boulders to the present places. Furthermore, Louis Agassiz, in his book Studies on Glaciers, wrote, "Thus the Swiss Alps became the center of the phenomenon of transportation of the erratic boulders scattered over the great Swissplains, the Jura and Northern Italy." Also, the theory said that the erratic boulders represent so many fragments, broken loose from the Alpine massif during its uplifting and that consequently, they could have been transported to their present location only after the uplift. We are naturally bound to inquire why they did not fill our lakes. There are only two alternatives: Either the lakes were protected in some way from the invasion of the boulders, or they did not exist at the time of the transportation. Similarly, northern Europe is the center of another region of erratic boulders, which are scattered over England, Germany, Poland, and Russia. Also, in North America, there are hundreds of boulders left from the Pleistocene times.

Further, the case of the boulder and how they got there raises many questions. For instance, the gradient from the Alps to the Jura is considered too low to allow the progression of a mass of ice like a glacier. Also, the slopes should be greater, for the flowing of water streams are capable of transporting erratic boulders. Louis Agassiz asked, "But how, did it happen that enormous amount of large pebbles and gigantic boulders scattered between the Alps and the Jura. On the Swiss plain as well as along the foot and on the slope of the Jura, did not fill them? How did it happen that their shores still display unequivocal traces of the abrasion and the polish no longer shown by their bed?" The erratic boulders occurred outside the Alps are scattered at different elevations, in the great Swiss plain at the foot of the Jura and at all elevations, on its southern slope, as well as in the internal valleys of that range. The famous Pierre-A-Bot boulder, which has a volume of about 50,000 cubic feet, occurs on that horizon at an elevation of 2,177 feet; there is a great boulder at an elevation of 2,772 feet on the northern slope of the Montagne de Boudry.

The shape of the erratic boulders of the Jura is also worth of attention because they are generally angular and devoid of traces of abrasion or friction. The transportation of these boulders from the Alps to the Jura has, at all times, puzzled geologists. Obviously, the agent responsible for it must be an extraordinary power. We can assume any natural power like icebergs or a great flood as probably responsible for this phenomenon but still beyond our human comprehension. The hypothesis of great currents has been, for a long time, widely accepted although the theory of currents is unable to explain all the features of the erratic boulders in the Alps and Jura. The great geologist Von Buch clearly saw the impossibility of the existence of a single current. The reason is that the boulders, instead of having been deposited at different elevations all along the Jura, should, on the contrary, have been accumulated in the direction of Jeneva. So Von Buch assumed on the basis of the petrographic variety of the erratic boulders in the different regions. "A number of currents equal to the number of distinct regions" he had recognized. He distinguished, in particular, the currents of Valais, Reuss, Aqr, and Limmatf. These currents received, at the place of origin of the boulders, "an extraordinary impulse capable of maintaining the boulders at distinct elevations inside the water during their transportation." Certainly, this theory should explain the difference in elevation between the boulders occurring in the plain and along the shores of the Lake Neuchâtel. However, this theory implies a more complicated circumstance that made it impossible to solve the mystery. Certainly, the impulse responsible for the assumed current should have removed, instantaneously and simultaneously, boulders originally located at very different elevations. Furthermore, the impulse would have had an unbelievable power in order to be able both to maintain these boulders, coming from different geological formations on their original path and to prevent them from being mixed in the middle of all the obstacles encountered by the current.

For example, the most accurate cannon cannot fire simultaneously several shells in a perfectly parallel direction, even at short distances. So my question is, how could an impulse less clearly defined be able to transport boulders in a mobile system as a water current? Additionally, how does this impulse in the water maintain the boulders in a parallel order and deposit them at different elevations? For instance, some boulders have been deposited at 5,100 feet in height. Also, another question is, why have these boulders not been abraded and rounded during transportation since they must have received mutual impact and hit the walls of the valleys they were crossing? Why has no size or density sorting taken place? Indeed, it has been said that these currents were swift enough to carry boulders as well as pebbles and that boulders never had a chance to hit the bottom or even to be rolled about. Another question is, why were the boulders not shattered when they hit the Jura? And why does the resistance of material play no role? Furthermore, why the presence of boulders on top of the highest ridges and their absence in the lower valleys?

Likewise, is well known that even currents of moderate speed, upon hitting rocks, generate whirlpools and eddies of great violence while dragging towards them every mobile object. How could currents with a volume and speed strong enough to transport boulders, could have been able to deposit them in their present locations in such a position? The eddies is a current moving contrary to the direction of a main current would have passed above without disturbing them? Or if eddies carried the boulders, why did they not drag the boulders toward the lower areas instead of leaving the boulders on ridges? Also, the existence of such a current, which is supposed to have flowed over Switzerland, is by itself a mystery. Does it really exist? Moreover, with respect to polished surfaces and striations in the middle of the Alps, they are considered as resulting from the action of water. The question is about the origin of the water, which could have had such a powerful effect on rocks, or where the reservoirs were located that have supplied currents for a rather long time and given them a powerful impulse to transport simultaneously boulders from all the ridges of the Alps in all directions and up to the summits of Jura? By the same token, upon reaching the Jura, these assumed currents should have flowed either eastward or westward and consequently should have built longitudinal streaks of boulders, but they are nowhere to be found. Or assuming that the water drained away, how could this be possible without the lakes remaining filled? Similarly, how can the theory of currents explain the angular shape of erratic boulders, which is the most striking feature?

C. Lyell offered the theory that the transport of angular boulders had taken place on top of ice sheets carried by water currents. My question is, how can ice lift boulders that weighs hundreds and thousands of tons to mountain tops? For instance, how the erratic boulders got on top of Mount Washington in New Hampshire at a high of 4,000 feet much higher than the source in Quebec from which the boulders are supposed to come. The great Canadian glaciologist A. P. Coleman writes, "The climber who spies a bit of foreign stone on the top of Katahdin, Washington or White Face should look on it with interest as evidence of strange events. It was picked up probably hundreds of thousands of years ago on the lowlands of the north, was lifted a mile in the grasp of a vast glacier and released on the summit of the mountains." Although a boulder thousands of tons heavy violates the laws of gravity, because the stone is heavier than ice. Also, the boulder would continually fall downward. Moreover, Lyell assumed that icebergs transferred boulders over the sea.

Also, he thought that the land was submerged and icebergs traveling over it dropped their loads of stones. Gradually the land emerged with the boulders in it because erratic boulders are found in the top of mountains. So he thought that the continents were under shallow water under recent times. Likewise, in some places, erratic boulders are distributed in a long string as in the Berkshire. However, icebergs could not act as intelligent carriers. Interestingly, some boulders lie on the Jura Mountains at an elevation of 2,000 feet above Lake Geneva. Some of these boulders are thousands of cubic feet in size, and a particular boulder named Pierre a Martin is over 10,000 cubic feet. They must have been carried across the space now occupied by lakes and lifted to the heights where they are found.

By the same token, in the British Islands, on the shore and in the highlands, are enormous quantities of boulders transported there across the North Sea from the mountains of Norway. An enormous force wrested them from those massifs, bore them over the entire distance that separates Scandinavia from the British Islands, and set them down on the coast and on the hills. Similarly, from Scandinavia, boulders were also carried to Germany and spread over the country. In fact, in some places, there are so many that it was thought they had been brought there by masons to build cities. Even in the Harz mountains in Germany, there are boulders that originated in Norway. Likewise, from Finland, boulders were carried to the Baltic regions and over Poland and lifted onto the Carpathians. In the United States, erratic boulders broke from the granite of Canada and Labrador and were spread over Maine, New Hampshire, Vermont, Massachusetts, Michigan, Wisconsin, and Ohio. The boulders perch on top of ridges and lie on slopes and deep in the valleys. Also, they lie on the coastal plains and on the White Mountains and the Berkshires, sometimes in unbroken chains. In the Poconos Mountains, they balance precariously on the edge crests.

Some boulders are enormous; for instance, the boulder in Conway, New Hampshire, is 90 × 40 × 38 and weighs 10,000 tons. Also, the Mohegan boulder in the town of Montville in Connecticut. Moreover, the great flat erratic boulder in Warren County, Ohio, weights approximately 13,500 tons and covers three-fourths of an acre. In addition, the Ototoks Erratics situated 30 miles south of Calgary, Alberta, in Canada consists of two pieces of quartzite brought from at least fifty miles to the west and weighs approximately 18,000 tons and is 250 × 300 feet in circumference. Similarly, in Malmo, Southern Sweden, there is a mass of chalkstone that is three miles long, one thousand feet wide, and from one hundred to two hundredfeet thick. The boulder has been transported a great distance, and a village has been built on top of it. All over the world as well, on isolated islands in the Atlantic, Pacific, and in the Antarctic lie boulders of foreign origin brought from afar by some great force. The boulders have been broken from mountain ridges and coastal cliffs and then carried down and uphill, over land and sea. After all, there is another theory that postulates that the boulders had been entangled in a tree or bushes and that is the way they traveled around the world. "When the roots of a tree encounter a big stone, they grow out and around and entwine it. So that when flood waters torrents rip out the tree the stone or boulder goes floating, over the sea with the lighter branches catching the breeze and sailing half a mile or so an hour, till the waves washes the trees ashore."

Obviously, it is ingenuous to think that ten-thousand-ton boulders are transported this way across oceans and continents. Also, how can we account for boulders buried in coal mines? Additionally, the scientist J. A. De Luc postulated the theory that erratic boulders were transported by air. So boulders had been projected through the air as far as the Jura by eruptions that occurred in the Alps, also by gaseous eruptions generated by the sinking of the layers forming the valleys. Although, the geologist de Saussure demonstrates the impossibility of projecting boulders through the air. He said, "Indeed, boulders of such enormous weight, coming from as far, as the central part of the Alps, and consequently following an extremely high trajectory, should have shattered the rocks on which they landed, and should have generated large depressions at the point of impact. Nonetheless, the boulders just lie gentile on the surface of the earth. Also, if they had fallen from an elevation of 8 or 10 feet, they would have generated cavities on the ground."

Obviously, that is not the case. Therefore, erratic boulders weighting 10,000 tons are not transported by ice, because the weight of the boulder would destroy the ice. The boulders cannot be transported by water or the Great Flood, because there has to be a supersonic speed by the water currents or the boulders would sink immediately. So the main question still is, how were the erratic boulders transported around the world across oceans, mountains, continents, and islands?

My theory is that the intelligence behind the UFO phenomenon is responsible for the transportation of the erratic boulders around the world. The intelligence behind the UFO phenomenon teleported the erratic boulders, which is the transportation of matter through space by converting it into energy and then reconverting it into energy and then reconverting it at the terminal point. The intelligence behind the UFO phenomenon teleported the erratic boulders and also teleported the Great Biblical Flood. The basis of teleportation is transferring information without sending it through ordinary space. It's a transfer achieved with a mysterious phenomenon called in science entanglement, a bizarre shifting between nature's tiniest particles no matter how far apart they are.

Entanglement lies at the heart of teleportation. For instance, I'll show in the next following cases how the UFO phenomenon teleported the boulders. In this case in England during the autumn of 1977, a series of shocking events took place at a place name Ripperston Farm owned by the Coombs family. The family reported to the police UFO sightings a couple of times. The first major event occurred on April 16. Pauline Coombs was driving home at night with three of her children. Then her ten-year-old son, Keiron, who was sitting in the backseat, reported a strange light in the sky. The light was luminous, yellowish with a grayish light underneath and a torch-like beam shining down from it. Keiron told his mother that the light had U-turned and was following them. Eventually the object caught up with the car and traveled alongside it. As a consequence, the car lights began to fade. Suddenly the UFO took off and vanished in the night sky. Moreover, on April 22, Mrs. and Mr. Coombs were watching a late-night movie on television, but there was a very strong interference. At about 11:30 p.m., Mrs. Coombs became aware of a glow outside the sitting room widow. An hour or so later, her husband saw a face at the window; he said, "It was a man, of a terrible size."

He estimated the height at seven feet tall, and he was wearing white suit. His face, if he had one, was concealed behind a black visor. Mr. Coombs was terrified and decided to call the police. The police came looked around but found nothing.

The most shocking event reported by the Coombs family was the vanishing and reappearing of the cattle. The Coombs family reported, sometimes one or two cows disappeared, but frequently the entire herd had disappeared from the farm! Sometimes he received an angry telephone call from a neighboring farmer asking him to come to collect his herd. Mrs. Coombs insisted that the animals had been properly fastened in, adding that Mr. Coombs had secured the bolt with binder twine as an extra precaution. According to Mr. and Mrs. Coombs, for the herd to escape to the next farm in that way indicated "the herd would have had to move past the cottage!" Yet neither he nor his wife had heard a sound! In addition, on another occasion, he reported there simply had not been enough time between the moment at which the cattle had last been seen and the moment when they were reported at another farm for them to have traversed the distance in any natural way. "The cattle were badly frightened, and the production of milk went down."

The implication of this supernatural event is that the herd of cattle or cows was teleported by the UFO phenomenon. Again, everything began with a UFO sighting by the Coombs family. In another farm in England, after a couple of UFO sightings, the workers and family reported that sometimes a cow would be found up in the hay loft. How it got there was a mystery as there was no passage large enough to admit an animal of such size. On another occasion a horse vanished from "the barn and was found in the hay room."

I believe in the same way the intelligence behind the UFO phenomenon teleported the boulders the water in the Flood was probably also teleported. So after the last Pleistocene mass extinction and the mess left on the planet, the intelligence behind the UFO phenomenon began creating monuments around the world using the same technology of teleporting great stones.

The first example is the great pyramids of Egypt; there are many questions. How were the pyramid stones dressed? Dressing many of these stones would require a pressure of two tons! How did primitive

Egyptians get the necessary equipment to apply that kind of pressure? How did they get the stones to the building site without cranes?

Furthermore, Stonehenge in England is another architectural mystery. Interestingly, the site of the quarry is 240 miles away and each stone weighs up to 5 tons. Stonehenge was built approximately between 1900 and 1600 BC, q thousand or so years after thepyramids of Egypt, a few years before the fall of Troy during the flourishing of the Minoan civilization. The prehistoric people that built them left no evidence of the way was built. The questions remain—how were these stones transported in a very irregular and hilly terrain? What kind of tools were used to quarry the stones?

Similar is the stone platform of Baalbek in Lebanon, which is situated in Beirut. This is a gigantic terrace composed of stone blocks, most of them with sides more than 60 feet long and weighing up to 2,000 tons and is situated at a height of 3,760 feet! Again the question is how and when it was built, and who built it? What kind of tools were used? Nobody knows.

By the same token, in Easter Island, there are statues that weight 50 to 70 tons. The hats in the statues are made of stone and have a volume of approximately 200 cubic feet and could weigh as much as 4 tons. How can the aborigines have lifted 4 tons to the level of the roof of a four-story house when there were no cranes and not even a high point in the vicinity? The quarry was a mile away, and how were hundreds of giant statues transported across the island over distances as long as eleven miles by people who lacked draft animals and wheels?

Likewise, the ruins of Sacsayhuamán in Peru—local archaeologists don't think the Incas built this fortress of stone. The fortress is situated at a height of 11,480 to 12,415 feet. The monolithic block weights more than 100 tons! The ramparts are 18 feet high. The terrace wall is 1,500 feet long and 55 feet high. Again the same questions—who created these massive stones? Because to have built such immense structure in a terrain so elevated and irregular would have been a fantastic feat that the poor Indians who lived in the area did not have the physical strength nor technology to create. My answer is that the intelligence behind the UFO phenomenon teleported these stones and boulders and built these great monuments to amaze the human race.

Odd Geological Mysteries

Furthermore, nature is full of mysteries that science cannot explain as a normal, natural process. For instance, the late, great scholar Immanuel Velikovsky mentioned in his masterpiece Earth in Upheaval a geological puzzle called the Carolina Bays. He wrote "'Peculiar elliptical depressions' or 'oval craters' locally called 'Bays' are thickly scattered over the Carina coast of the United States, and the entire Atlantic coastal plain from southern New Jersey to northeastern Florida. The marshy depressions are numbered in the thousands and according to an estimate, their number may reach half a million." Moreover, the Geological Society of America in 1952 gave the measurements made on more prominent ones seaward from Darlington. They show that the larger bays average 2,200 feet (twothird of a kilometer) in length and in single cases exceed 8,000 feet (over 2½ kilometers or more than a mile a half).

'A remarkable feature of these depressions is their parallelism. "The axis of each of them extends from northwest to southeast, the precision of the parallelism is striking. Additionally, around the Bays are rims of earth, invariably elevated at the southeastern end. These oval depressions may be seen especially well in aerial photographs. Any theory of their origin must explain their form!"

Interesting are the ellipticity, which increases with the size of the Bays; their parallel alignment; and the elevated rims at the southeastern ends. The origin of the bays apparently cannot be explained by the current geological theories or activity of the earth. An extraordinary process must have created these geological puzzles. Such a process is suggested by the elliptical shape, the parallel alignment, and the systematic arrangement of elevated rims. I postulated my theory that the intelligence behind the UFO phenomenon is responsible for these geological formations. As a matter of fact, during the Tunguska explosion in 1908, a UFO or something detonated by the UFO left elliptical marks in the burned forest. Ingenuously, some geologists believed that meteorites, asteroids, comets, and shooting stars made those elliptical marks. Also, geologists think that the catastrophe took place sometime during the last Pleistocene mass extinction.

Another great mystery is the Rock of Gibraltar, which seems upside down. The story of Gibraltar is complex and puzzling because it contradicts every established geological theory. The principle established by geologists is that younger sedimentary rocks are superimposed upon older ones and that when deposited, they spread, laterally and predictably, in all directions to form distinct layers. How are rock layers formed? Actually, since the mid seventeenth century, a theory of geology was postulated by Nicholas Steno. He wrote that there were three fundamental principles that govern the formation of sedimentary strata.

His first rule is about the principle of superposition and is the most obvious. He wrote,

> "In any sequence of undisturbed sedimentary rock strata. The oldest rocks are on the bottom, with the younger ones stacked successively atop them. Even the most casual inspection of the rocks making up a cliff or hill-side shows that they exist in layers. Undoubtedly, it seems implicit that the layers of different rocks types within the earth's surface, represents A time sequence. Unless something extraordinary has happened! The deeper rocks are older; closer to the surface the rocks are younger! and that applies also to the fossils they contain. Also, geologists stated that if one layer of rock lies on top of another, then the lower one was deposited first and played no part in the formation of the upper one. In addition, geologists said that all beds of rock were originally laid down horizontally. Similarly, the beds of rock are laterally continuous until they are replaced by another bed of the same age. The implication is that the layered rocks represent a time of sequence and that the rocks exposed at any one small locality were not a separated set of structures resulting from unique

processes, but were in principle related to other exposures. In the case of the Rock of Gibraltar the currents theories just mention, do not explain the order of the layers in which the Rock of Gibraltar is. For instance, the Eocene strata is young in the geological context, which are covered with older Upper Jurassic, which in turn is covered by still older Lower Jurassic. Obviously, this is a complete reversal of the accepted geological evolutionary sequence. The order should be: First Lower Jurassic, then Upper Jurassic, with youthful Eocene on top. In fact, any way to explain this phenomenon as having washed away by erosion, or possibly as having "drifted" away on a convection current don't explain anything. As a matter of fact, a geologist explained the only way for this event to take place is: "To slide the Jurassic nappes in, on top of Gibraltar, as Holmes suggests from the east by uplifting the Mediterranean floor, which is now at a depth of 2,000 metres. Or else invent a theory that will turn Gibraltar upside-down in order to set its rocks right-side up." In the same way Mount Blanc was moved from its place and the Matterhorn was overturned. Likewise, the Alps like the Rock of Gibraltar is in a confused geological sequence. For example, the Cretaceous and Jurassic strata are on top of Eocene strata, instead of being underneath as the principles of geology demands. Moreover, the Bahariya geological formation is another puzzle like the Rock of Gibraltar."

The geological strata are not in sequence. In fact, it has been known for more than a century that almost all the layers in the Bahariya formation were formed during what is known as the Cenomanian Age at the very beginning of the late Cretaceous, which began 99 million years ago. Also, it has long been understood that the various bands of rock that make up the slopes of Gebel El Dist are neatly stacked and firmly compressed layers of sediments that were deposited in shallow estuaries and in saltwater lagoons over a few million years, during which time the Tethys Sea inundated the land and then retreated from it. Moreover, it has also been known that the limestone cap that sits atop these late Cretaceous sediments was formed during a much more recent period, the Eocene epoch of the Cenozoic Era, which followed the Mesozoic Era. The puzzle is that the Eocene epoch did not follow directly after the late Cretaceous; it did not even begin until 37 million years after the rock in El Dist was formed.

I postulated my theory that the intelligence behind the UFO phenomenon used antigravitational technology, and that changed the time and space; as a consequence, it changed the age of the strata. Interestingly, Albert Einstein said once that time was an illusion; probably he was right that we live in an eternal present. So the intelligence behind the UFO phenomenon can distort our time and space with anti-gravitational technology.

Sinking and Raising Continents

Interestingly, there is a mystery of the raising and sinking continents. For instance, there are the famous myths about the continent of Atlantis, Lemuria, and few others. Probably, in this discovery by a Columbia University geophysicist, studies can help us to find clues about this mystery. In the 1950s, a Soviet scientific research discovered the Gamburtsev Mountains buried beneath two miles of ice in East Antarctica. The radar waves bounced off the rock beneath the ice; returning echoes showed the contour of peaks and valleys as well as hundreds of miles of interconnected subglacial lake systems.

After studying the results of the survey, the scientists were stunned by what they saw. "The range of 9,000-foot tall mountains, mighty as the Pyreness, seemed to have sprung up out of nowhere. Normally mountains arise where you've smashed continents together or from volcanoes of subduction zones. But none of those exciting things seemed to have happened under Antarctica for hundreds of millions of years, that was the mystery."

Indeed, this geological event gives us the clue about what happens to continents that sink in the ocean. Conversely, if mountain ranges seem to appear without any of the geological laws that we know, surely continents sink and disappear without any of the geological laws we know. Obviously, if there are no natural process involved, something causes this phenomenon and must be supernatural. After all, the only

ESSAY ON THE THEORY OF THE EARTH: ELECTROMAGNETISM IN UFOS AND THE ORIGIN OF MASS EXTINCTIONS AND THE ICE AGES

supernatural cause that I can think of is the UFO phenomenon. Geologists can't explain the sinking of the sediments through which low-density sediments apparently displace higher-density rocks. Geologists have referred to the formation of mountains and geosynclines as one of the major unsolved mysteries of geology. Moreover, geologists stated, "Low density sediments apparently displaced higher density rocks, is heightened when we note that these narrow elongated zones in the earth's crust, down warped the most, with the greatest accumulation of rock debris, shed by ranges and the highest portion of the continents."

Some geologists postulate complicated mechanisms to explain this phenomenon; they stated, "The most puzzling aspect of the matter, is that the trough, does not always seen to be formed in advance of the filling, but to deepen gradually as sediments accumulates in it. Undoubtedly, the most mysterious is that the mountain building movements are by spasms. It means that the sediment went down gradually, but came up by spasms. It went down presumably, because it must have been heavier than the surrounding rock that it displaced. But what made it come up again? And why by spasms!"

So modern geology can't explain why continents sink or rise. According to geologists all over the world, they are in wrong order, or ages are missing! In Jamaica; the greater part of Mexico; the Athabasca district; Manitoba, Tennessee; Russia; the Island of Timor; China; California; Idaho; Utah; North and South Dakota; Greece, parts of South America, etc., shockingly, there is no place in the entire world where the rocks are in the right order. Although, there are places in the world where the geological strata are in the proper order. Sir Charles Lyell wrote, "Violations of continuity are so common, as to constitute even in districts of considerable area, the rule rather than the exception." In addition, geologists stated that there is a sort of frequency modulation, as it were, during the carboniferous age.

Eastern United States, or at least the Pennsylvania part of it, has risen and fallen about forty-two times. Similarly, according to geologists, there is one complete cycle per half-million years. On the other hand, parts of England and Germany oscillate sixty and thirty-three times respectively. Some geologists postulated the theory that during the Coal Age the rock crust oscillated once in five million years. Also, they think that it is odd that the rock crust oscillated with such high frequency in the Coal Age. Geologist J. Marvin postulated the theory that forty-two uplifts and forty-two subsidence had taken place during the Coal Age in the Pennsylvanian period. Each cycle covers a wide area and involves an advance, a retreat, and a re-advance of the sea.

Further, geologists think in the absurdity of all these sinking and rising in the United States as well as all over the world. All these phenomena need an explanation and revision of our current geological theories. In fact, every time the rock sinks, it must displace the underlying rock and push it to each side. As a consequence, every time the rock rises, the rock on each side must flow in again to fill the space. Although, some geologists think that something wrong is with this theory because solid rock does not move up and down, and still less does the rock on either side move in and out. In fact, the whole theory of geology must be reconstructed. The late great scholar Charles Hapgood thought that it may be erosion that causes the up and down of the rock strata. But erosion cannot wear down rocks below sea level for the simple reason that rivers cannot carry sediments uphill; as a consequence, we need to invent some kind of force that will pull the rocks down from below. \

A geologist explains, "Denudation can lower a region but it cannot bring it below sea level, and hence land areas that founder into the sea, must also have been pulled down by some force from the interior." However, there are, many times, more push-ups than there are pull-downs. Certainly, there is no theory that can explain why the rocks' strata rise more frequently than they sink. The college textbooks in geology explain uplifts and subsidence of the land with the theory of isostasy and, according to this theory, light rock rises and heavy rock sinks. Interestingly, the textbooks on geology seldom, if ever, tell the students that there are events of negative isostasy that poke holes in the theory. For example, North America has been no less than seventeen times underwater, or in seventeen times, the heavier rock of the continent sank into the lighter rock beneath, displacing it to each side. Moreover, on seventeen occasions, the lighter rock of the continent routed out the heavier, allowing it to flow underneath, though the principle of isostasy does not explain what make the rock lighter or when the theory requires that it should rise. Likewise, what changes the density and makes it heavier, when the theory requires that the rock should sink.

My theory is that the UFO phenomenon, after reversing the magnetic fields and reversing the magma, can cause these changes in the crust of the earth. As a result of the expanding of the ocean floor, they can cause changes in the crust of the earth, causing the sink of the continents or the rise of mountains and islands. Although, my theory is based in the possibility that the earth changes are created by the UFO phenomenon.

Plate Tectonics

Alfred Wegener, a German meteorologist in 1915, published his theory of plate tectonics. He noticed that the coastlines of Africa and South America looked as though they would fit neatly together if the Atlantic Ocean were removed. He explained his theory in the book The Origin of Continents and Oceans. Interestingly, prominent geologists and scientists of the day didn't take his theory seriously. The scientists believed that the outer part of the earth was "much too rigid to permit the continents to drift about like ships on the sea." The scholars said that what Wegener thought was the drifting of the continents was caused by the centrifugal forces resulting from the earth's rotation. On the other hand, the scientists of the day thought that the forces created by the centrifugal forces were too weak to cause the continents to drift.

As a matter of fact, Wegener's theory lacked a mechanism to cause the continents to drift. Geologists said, "Without a driving force, continental drift can't occur." Nevertheless, continental drift is a theory accepted by modern science, and it is accepted that Africa and South America were joined in a supercontinent. According to modern geologists, at least once in the earth's history, all the present continents were linked to form a supercontinent that stretched from pole to pole. In modern times, we called them plate tectonics, and the source of movement of the plate tectonics is the magnetic properties of the sea floor.

Moreover, scientists have discovered that magnetic stripes on the sea floor, magnetic reversals, and continental drift are all interconnected. Scientists discovered that the zebra-stripe magnetic pattern on the seafloor records exactly the same sequence of magnetic reversals as the continental basalts. Also, scientists agree that seafloor spreading is a reality and new ocean crust is being formed by lava continually welling up at the center of the ridges. The magnetic pattern is symmetrical because the lava is magnetized, and as it cools off to solid rock, it spreads away equally to both sides. The seafloor acts as a kind of gigantic magnetic tape recorder, recording the reversals of earth's magnetic field. Further, the dates of the reversals are known from analysis of rocks on land, and the magnetic stripes on the ocean floor can be used as time markers. The rate at which new seafloor is being created can be calculated by measuring the distance from the center of the ridge, where the age of the seafloor is zero.

Although it varies from place to place, the rate of seafloor creation is calculated as several centimeters per year. In fact, this is about the same speed at which your fingernails grow. The continents on either side of the Atlantic are moving apart at this rate. The entire Atlantic Ocean can be created in less than 200 million years, which is not too old in geologic terms. In fact, no seafloor in any of the world's oceans is much older than 200 million years old. Obviously, the continents are well attached to the rocks of the seafloor, and they move apart at a velocity that is governed by the rate of creation of new seafloor at the mid-Atlantic ridge.

The scientists' objection to Wegener's theory of continental drift is that the continents don't plow through the rigid rocks of the ocean floor. As a matter of fact, what takes place is that continents and ocean floor move together, both part of a lithospheric plate. So the intelligence behind the UFO phenomenon when it reverses the magnetic field also is able to move the seafloor and sink continents or create new mountains. I think it is through the control of the magnetic fields that the UFO phenomenon can control the other phenomena like tectonic plates, the sinking of continents, and rise of mountains without continent collision.

The earth's crust consists of a number of separate plates—ten major ones, subdivided into varying sizes made of rock, 40 to 60 miles thick, which float on the hot, viscous mantle beneath them. The earth's land surface, which rests on these plates, as do the ocean floors, was once welded together in a single supercontinent. Approximately 200 million years ago, the continents began to split up. Eventually the present seven continents and major islands were formed. The continents have slowly drifted apart like packages in a conveyor belt. The crustal plates are being built up at their edges by molten rock welling from deep fissures in mid-ocean. At the same time, these plates are being propelled across the globe by forces arising from deep within the earth in various directions at the speed of one-half inch to six inches a year.

Moreover, when a moving continental plate (mainly granite) meets an ocean floor plate (dense, less buoyant basalt) the continent rides over the seafloor plate. It scrapes off the layers of sediments deposited on the seafloor over many millions of years, as well as slices of crustal rock. This debris piles up along the edge of the continent like a rumpled blanket and forms mountain ranges. Interestingly, no one could conceive of any mechanism that could propel vast continents through the earth's solid crust. Moreover, it was thought that such moving land masses would have left behind gigantic wakes of displaced rock on the seafloors. Even through the scientists' intensive searches, no ripple of such disturbance was discovered.

Another mystery is the strange sparsity of sediment on the ocean floors. Geologists had calculated that the sediment formed by microscopic marine organisms, together with dust blown or washed into the sea, should have blanketed the ocean beds over the ages to a uniform depth of at least twelve miles. Yet scientists found no sediments in the center of the world's oceans and only a half mile thick near the border adjacent to the continents. Obviously, it is unacceptable, crazy, or out of our minds to even postulate a theory that considers the possibility that an alien intelligence is involved in nature's processes here on earth. I see the possibility that the intelligence behind the UFO phenomenon is responsible for keeping or charging the magnetic fields, reversing the magnetic fields, and causing the greatest mass extinctions and natural disasters like the Flood, the Ice Ages, etc. As far as I know, the only supernatural agency operating on Planet Earth is the intelligence behind the UFO phenomenon. Also, the UFO phenomenon is responsible for the teleportation of boulders across the world.

CHAPTER 4

UFOs, Asteroids, Comets, and Mass Extinctions

> The disturbing reality is that for none of the thousands of well-documented extinctions in the geologic past, do we have a solid explanation of why the extinction occurred. We have many proposals in specific cases, of course: Trilobites died out because of competition from newly evolved fish; Dinosaurs were too big or too stupid; the Antlers of Irish Elk became too cumbersome. These are all plausible scenarios, but no matter how plausible, they cannot be shown to be true beyond reasonable doubt. Equally plausible alternatives scenarios can be invented with easy, and none has predictive power in the sense that it can show a priory that a given species or anatomical type was destined to go extinct.
> —David M. Raup

The Explosion of June 30, 1908, in the Tunguska Forest in Siberia

In 1921, the new Soviet Academy of Sciences commissioned a remarkable scientist named Leonid Kulik to collect information about the explosion in the Tunguska forest in Siberia. The local newspapers of Irkutsk, Tomsk, and Krasnoyarsk had all reported the event. The newspapers described it as "the most unusual phenomenon of nature." In the village of Nizhne-Karelinsk in the northwest high above the horizon, the peasants described a shining body, very bright (too bright for the naked eye) with a bluish-white light. It moved vertically downwards for about ten minutes. The body was in the form of a pipe (cylindrical).

More peasants reported, "The sky was cloudless, except that low down, on the horizon in the direction in which the glowing body was observed, a small dark cloud was noticed. It was hot dry and when the shining body approached the ground it seemed to be pulverized and in its place a huge cloud of black smoke was formed and a loud crash, not like thunder, but as if from the fall of large stones, or from gun fire was heard. All the buildings shook and at the same time, a forked tongue of flame broke through the cloud." The peasants wept, because they thought that the end of the world was approaching.

Moreover, the village of Nizhne-Karelinsk was situated at about 200 miles (320 km) away from the centre of the explosion. According to a local meteorologist named Voznesensky, the explosion had been heard 500 miles (800 km) away from its center, and at that distance, the seismic instruments in Irkutsk had registered a crash of earthquake proportions. The peasants reported a "fiery, heavenly body, a flame that cut the sky in two and a pillar of smoke."

Another account was by a peasant named Ilya Petapovich when she was going to the spring for water. She said, "One day a terrible explosion occurred, the force of which was so great that the forest was flattened to the ground, its roof was carried away by the wind. And his reindeer fled in fright. The noise deafened my brother and the shock caused him to suffer a long illness."

Another peasant named Vasiley Okhchen mentioned how he and his family were asleep when, together with their tent, the whole family went flying into the air. "All the family were bruised but Akulina and Ivan lost consciousness. The ground shook an incredible long roaring was heard, everything round about was surrounded in smoke and fog from burning falling trees. Eventually the roar died away, but the forest went on burning. We set off in search of the reindeer, which had run away. Many did not come back."

Another peasant from the village of Vanavara reported, "I saw the sky in the north open to the ground and fire pour out. We thought that stones were falling from the sky and rushed off in terror leaving our pail by the spring. When we reached the house, we saw my father unconscious lying near the barn. The fire was brighter than the Sun, during the bangs, the Earth and the huts trembled greatly and earth came sprinkling down from the roofs."

Moreover, there were stories of horses bolting with their plough, and a man felt his shirt burning on his back. Many wildlife like reindeer and domestic dogs died. Also, many houses were destroyed. The scientist Leonid Kulik, who was sent by the Russian Academy of Science, reported, "Trunks of the falling pines, they all lay with their tops facing uniformly towards the southeast." He climbed the heights of the ridges and he saw "stretching as far as he could see, at least 12-16 miles (20-25 km) was utter desolation. The huge trees of the taiga lay flat, firs, pines, deciduous trees; all had succumbed. The sharp outlines of the winter landscape etched it like a plate and again, this bizarre and unbroken regimentation." Further, despite the destruction, the trees still lay in only one direction. Whatever had caused the Tunguska explosion, Kulik thought that it had destroyed 37 miles (60 km) in one direction alone; this was the epicenter.

Interestingly, the most careful search of the epicenter and nearby terrain did not produce a sign of a meteorite impact. Through the aerial survey of 1938 and the close examination of the trees, their burn marks and the pattern of their fall gave the scientists some idea of what happened during the explosion and the true size of the devastation, approximately 770 square miles. The size was as big as Birmingham in England or Philadelphia in the United States. Strangely there were odd features in the area of the explosion like right in the middle, a large number of trees were left standing, though stripped of their branches. After all the surveying, digging, and searching, there were no signs of any impact had hit the Earth!

After the explosion, the peasants went to look for the reindeer and they found their charred reindeer carcasses. A peasant reported, "I was sitting on my porch facing north, when suddenly, to the northwest, there appeared a great flash of light. There was so much heat that my short burned off my back. I saw a huge fireball that covered an enormous part of the sky. Afterward it became dark and at the same time I felt an explosion that threw me several feet from the porch. I lost consciousness." As a matter of fact, this witness was forty miles away from the Siberian explosion.

In Central Russia, a deafening roar terrified the inhabitants of small towns and villages. Obviously, a powerful ballistic wave pushed before the descending object, trees were leveled, peasants' huts were blown away, and men and animals were scattered like paper. Shockingly, the explosion was of such force that the seismographic center at Irkutsk, 550 miles to the south, registered tremors of earthquake proportions. The vibrations traveled 3,000 miles through the ground to Moscow, St. Petersburg, and the earthquake observatory at Jena, Germany, 3,240 miles away also recorded strong seismic waves. Even as far away as Washington, DC, and Java, seismographs were activated by the immense blast. Immediately a gigantic pillar of fire appeared in the sky, and in towns 500 miles away, people heard a series of thunderous claps. The noise was so great that some peasants closer to the explosion were deafened. Others were thrown in a state of dazed shock that made them speechless. Further, with the brilliant fire in the sky, a searing thermal current swept across the forest, scorching the conifers and igniting fires that would continue to burn for days. The heat was felt by people forty miles away. At a distance of 375 miles to the southwest, hurricane-like gusts rattled windows and doors, collapsing ceilings, shattering windows, and flinging people into the air.

In Kansk, a station town with a newly completed Trans-Siberian railway, people walking or on rafts were hurled into the river. Similarly, farther south, horses stumbled and fell to the ground. Also, passengers in the Trans-Siberian Express were frightened by loud noises and were jolted off their seats. The train itself was shacked wildly in the tracks.

After, a dark mass of thick clouds rose to an altitude of more than twelve miles above the Tunguska region. Right after, the entire area was showered by an "ominous black rain." Furthermore, intermittent rumblings of "thunder resembling heavy artillery, reverberate[d] throughout central Russia."

Five hours after the blast, the air waves created by the object traveled west beyond the North Sea, causing strong oscillations at meteorological stations in England. Likewise, within twenty minutes, sudden fluctuations in atmospheric pressure were detected by newly invented self-recording barographs at six stations between Cambridge, fifty miles north of London, and Petersfield, sixty-five miles south of London. Baffled meteorologists thought that a large atmospheric disturbance had occurred somewhere in the world. Actually, after two decades, scientists discovered that their 1908 barographic records corresponded with the cataclysmic explosion of June 30, 1908, in the Siberian Tunguska forest. Moreover, it seems that the airwaves circled the Earth twice.

Also astonishing was the effect on the Earth's magnetic fields. In the 1960s, D. R. Vasilieyen pointed out "the evidence of electromagnetic chaos at the center of the explosion." Also, he said there was an "electromagnetic hurricane of enormous proportions which has shattered, perhaps permanently, all the normal alignments with the Earth's magnetic field."

Furthermore, Russian researchers noticed the similarity between the destruction at Hiroshima and the Tunguska explosion because there had been at least two blast waves and extensive fires and flash burns. Also, the scientists noticed an accelerated growth of trees and plants. The first American observers noticed that right in the center of the explosion, it had little damage and some trees remained upright, similar to the nuclear explosion at Hiroshima. Also at Hiroshima, the trees seemed to grow rapidly.

The similarities between Hiroshima and the Tunguska explosion are too close to ignore, although it is beyond belief to think that an atomic explosion had occurred in Siberia. The first atomic explosion was carried out by the United States forty years after at Almagordo. Dr. Vasilieyen of Tomsk University said, "There had been the most violent genetic changes, not only in plants but in the small insect life. There are ants and other insects quite unlike anywhere else. Some of the trees and plants just stopped growing, others have grown many times faster than before 1908."

By the same token, scientists found more similarities between the nuclear tests carried out by the British, Americans, and Russians, and the Tunguska explosion in Siberia. The Tunguska explosion produced extraordinary bright aurora lights and disturbances in the ionosphere. Dr. Vasilieyen stated, "It is certainly odd, I know of no other phenomena than the nuclear explosions which produce these displays at their magnetic opposite side of the world, though it could just be coincidence."

Another important feature of this explosion is that like in Hiroshima, the object that exploded at the Tunguska exploded in the air. Likewise, after the explosion, a mass of luminous silvery clouds appeared in Russia and Northern Europe. The light was so intense that during the next few nights, people took photographs at midnight and ships could be seen for miles out at sea. A Russian scientist describes the phenomenon as "a thick layer of glowing clouds as lit by some kind of yellowish green light that sometimes changed to a rosy hue. It was the first time I had seen such a phenomenon."

Additionally, extraordinary dust clouds and eerie nocturnal displays are observed for weeks across the continents as far south as Spain. Interestingly, on June 30, the day of the explosion, a scientist in Holland reported, "An undulating mass passing across the northwest horizon, it was not cloud, for the blue sky itself seemed to undulate." By the same token, after sunset in Antwerp, the northern horizon appeared to be on fire.

Conversely, in 1930, in the Royal Meteorological Society Quarterly Journal, a scientist gives an account of the odd colors he observed over England in 1908 on the nights of June 30 and July 1, 1908. "A strong orange-yellow light became visible in the north and north-east causing an undue prolongation of twilight lasting to day break. Twilight on both of these nights was prolonged to day break and there was no real darkness." The phenomenon was reported from various places in the United Kingdom, Copenhagen, Königsberg, Berlin, and Vienna. According to the London Times of July 4, 1908, "The remarkable ruddy glows, which have been seen on many nights lately have attracted much attention, and have been seen over an area extending over an area as far as Berlin." Witnesses reported that "abnormal glows appear in the sky only after the fading of twilight. The sky grows partially dark and then brightens again with deep, lurid color." A witness in London reported, "It was possible to read large print indoors and the hands of the clock in my room were quite distinct. An hour later, at about 1:30 a.m., the room was quite light, as if it had been day." Similarly, more witnesses reported, "The northern sky at midnight became light blue, as if dawn were breaking and the clouds were touched with pink in so marked a fashion that police headquarters was run up by several people who believe that a big fire was raging in the north of London."

In the London suburbs, people were drawn into the streets to view the frightening cosmic phenomenon. A meteorologist said, "I have never at night time seen anything the least like this in England, and it would be interesting if anyone would explain the cause of so unusual sight."

The Explosion in the Tunguska Forest in Siberia in June 30, 1908, and the Nuclear Blast in Hiroshima

The similarity—in the Hiroshima nuclear blast, wood was ignited at a distance of one mile. On the other hand, in the Tunguska explosion, wood was ignited at a distance of eight to ten miles. Further, from the fall point in Hiroshima, the blast destroyed an area of approximately eighteen square miles. At the Tunguska explosion, from the fall point, the area destroyed was 1,200 miles. In Hiroshima, people felt the heat 60 miles away from the blast. At the Tunguska explosion, people felt the heat 315 miles from the blast.

Furthermore, Russian and American scientists thought of the possibility that the Tunguska explosion was "100 times more massive in the megaton range." The Nobel Prize winner chemist Willard F. Libby has estimated its energy yield as high as 30 megatons or 1,500 times as great as that at Hiroshima. Moreover, in a 1966 report, Russians scientists determined that the magnetic and barographic effects as well as the zone of destruction of the Siberian forest were similar with the 1958 high-altitude nuclear tests conducted by the United States. Interestingly, if the Tunguska explosion had taken place in 1958, the advanced instruments at observatories would have detected a high-altitude atomic explosion had occurred. Likewise, some scientists were able to calculate the height at which the object had exploded in the air at the Tunguska explosion. Scientists thought that the object had exploded about 5 miles (8 km) up in the air. But what could have caused the appearance of a nuclear blast before the birth of the atomic age? Also, scientists though that the object weighed hundreds of thousands of tons.

Witness Accounts of the Object That Exploded at the Tunguska Forest

A newspaper published in Irkutsk, a village situated 550 miles from the explosion, reported that on the morning on June 30, 1908, Essay on the Theory of the Earth: Electromagnetism in UFOs and the Origin of Mass Extinctions and the Ice Ages89 89in a village north of Kirensk, peasants describe "a body shining very brightly. The villagers had dashed out into the streets in absolute panic; some wept in terror convinced that this must be the end of the world." The peasants stated, "And the pipe shape did not sound like a normal meteoritic object, the cloud of black smoke and the flame were also baffling."

Certainly reports of an enormous fiery object had been seen over villages and towns throughout the Yenisei river province. Some peasants described the object as almost moving horizontally from the south, and everybody had felt and heard Earth tremors and loud explosions. Peasants reported, "A subterranean crash and roar as from distant firing, doors, windows and lamps were all shaken. Five to seven minutes later, a second crash followed louder than the first, accompanied by a similar roar and followed after a brief interval by yet another crash."

Leonid Kulik, the scientist sent by the Russian Academy of Sciences, thought that it must have been a giant asteroid, and as a result, a giant crater must have been seen in the ground. But where had it fallen? The comet and meteorite theories found no explanation for the seismic shocks that registered around the world on June 30, 1908. Science had never noted shocks or earthquake-like tremors in the case of an asteroid striking the Earth. Nor had meteors ever disturbed the Earth's magnetic fields or gravitational field. Scientific research in the 1960s failed to find any electromagnetic disturbances caused by meteorites approaching the magnitude of the Tunguska explosion of June 30, 1908 in the Siberian forest. Interestingly, even if that technology in earthquakes was just beginning to be created, the primitive instruments of the day were able to register the worldwide tremors.

According to newspaper reports, the object had been observed over an area of more than 500 miles. From a Krasnoyarsk newspaper of 1908, which reported that in several villages along the Angara River, in the center of the taiga, peasants saw "a heavenly body of fiery appearance cut across the sky from south to north. When the flying object touched the horizon, a huge flame shot up that cut the sky in two. The glow was so strong that it was reflected in rooms whose widows faced north."

Furthermore, in other villages, horses and cows began to whinny and run wildly. A peasant said, "One had the impression that the Earth was just about to gape open and everything would be swallowed up in the abyss." Similarly, in the Trans-Siberian Railway, the station agent reported that he had felt a strong vibration in the air and heard a loud rumbling sound. The locomotive engineer had become so frightened by the ground tremors and noise that he had halted his train, fearing it might be derailed. Further investigators had found no signs of a meteorite or asteroid. Peasants mention a "huge area of flattened forest." Another witness reported "an unbelievable loud and continuous thunder; the ground shook burning trees fell and all around there was smoke. Soon the thunder stopped, the wind ceased, but the forest continued to burn." In addition, peasants reported that the "reindeer had scabs that never appeared before the fire came."

Obviously, what the peasants reported was a UFO; their description is similar to other descriptions given throughout history. For instance, in Russia, peasants reported on August 15, 1663, that between 10:00 a.m. and noon, "A great noise resound over Robozerd lake; from the north out of a clear sky appeared a huge flaming sphere not less than 130 feet in diameter, emitted two flame beams about seven feet in length; from its sides poured bluish smoke. This huge ball of fire hovered over the lake, after the great flame and two smaller ones vanished." The phenomenon was observed by dozens of witnesses, and some peasants were on boats in the lake, but the scorching heat forced them out of the water. Also, the peasants reported that "the light from the object, had penetrated the water and reached the bottom of the lake and the fish fleeing from the flame toward the shore."

Furthermore, one of the oldest reports in Russia on UFOs goes back to the year 1028. In the Russian Chronicles, there is a report of a sighting of a serpent-like object that could be seen throughout Russia. The object hovered for two days in a fiery pillar. It appeared from the ground, accompanied by thunderous noise. The witnesses thought that it was a sign from God.

Likewise, in 1317, over the city of Tuer, a UFO was seen for over a week. Similarly, on April 9, 1628, in Berkshire, England, a witness reported, "The weather was warm and suddenly a hideous rumbling was heard in the air followed by a strange and fearful pearl-like object and thunder. It sounded like a rough battle, a great cannon seemed to roar, then it takes a second time, till two cannon shots seemed to have been discharged in the sky. Then there was a sound heard like the beat of a drum sounding a retreat, then a hissing sound."

Similarly, on December 5, 1735, at 5:00 p.m. in England, a witness reported the appearance of "a deep red cloud under which a luminous body, sent out streams of very bright light, by which the witness could easily read shockingly, the streams of light moved slowly for some time and stood still!" The witness proceeded that "it became so hot that I had to strip my shirt in the open air. It looked like a great ball of fire. The object shook the ground with an explosion and all the sky seemed on fire." Likewise, on December 11, 1741, a witness reported the sighting of a ball of fire. A most terrible clap of thunder was then heard in the north, like two very large cannons fired a second after each other, but the rolling and the echoing were not like cannon shots. All the houses were shaken twenty miles around.

Moreover, in the twentieth century, one of the most important UFO sightings, the object reported looked just like the one reported during the Tunguska explosion on June 30, 1908, in the Siberian forest. On June 4, 1965, the American astronaut James McDivitt, during his Gemini IV flight, saw a cylindrical object ahead of their spacecraft. After the flight, McDivitt checked the records, and there were no rockets near the Gemini capsule at the time of the sighting.

Similarly, Gordon Cooper and Charles Conrad, during the flight aboard Gemini V on the third day of their mission, as they were passing Cape Canaveral (at that time it was called Cape Kennedy). The Gemini flight director, Christopher Kraft, asked them if they could see anything flying alongside the flight. The director radioed to the astronauts, "We have a radar image of a space object going right along with you from two thousand to ten thousand yards away." The object radar returned it was approximately the same magnitude as Gemini V. Interestingly, the object was tracked until both the object and Gemini V were lost beyond the curvature of the Earth. So what we see is a great similarity between the UFOs described in these events and the UFO described by the peasants during the Tunguska explosion in Siberia in June 1908. Here are the similarities between UFOs and the object that exploded at the Tunguska forest:

The Tunguska Object

1. Witness reported thunderclaps.
2. The object had a form of pipe, tube, or huge cylinder.

3. The object glowed bluish white radiance.
4. The tube-shaped object slowly descended.
5. The object disappeared in a flash of searing light.
6. The object produced earthquakes, electromagnetic heat, and radiation disturbances.

UFO Description

1. Witness reported thunderclaps.
2. Some UFOs look like tubes or cylinders like the one reported by James McDivitt during Gemini IV on August 24, 1965.
3. UFOs look like flaming objects.
4. UFOs appeared and disappeared in a flash of searing light.
5. UFO produced heat, electromagnetism, radiation disturbances, and earthquake-like tremors.

The Flying Path of the Tunguska Object According to Russian Scientists

The 1908 object's path of flight and the probable location of the blast had been estimated in the mid-1920s by a Russian scientist, A. V. Voznesensky, former head of the Irkutsk Observatory. The scientist used some information acquired by Kulik and Obruchen and earlier seismic data from Iskutsk and other Russian stations and observations of acoustical phenomena throughout Central Siberia. Voznesensky attempted to trace the path of the object and find the location of impact. He found that the effects of the explosion "had been heard by people over an incredibly immense geographical area. One larger than France and Germany combined." The "fiery object" racing through the cloudless sky had been observed by thousands from the southern border of Siberia to the Tunguska region. Similarly, the noise of the explosion, the thunderclaps, and rumblings like thunder were heard for a radius of 500 miles. From the reports and seismic data, he was able to figure out the time of the blast at about 7:17 a.m. on June 30, 1908. The place of the fall, he thought, was in the territory north of Vanavara in Siberia. Moreover, in February 1927, scientists departed from Leningrad with a research team to Kansk and farther east to the remote station of Taishet, and they were agreed on the northward direction of the object and the thunderclaps heard five hundred miles from the explosion.

So scientists, during their investigation, found, "The uprooted trees showed their tops always pointing toward the south. The direction in which they had been heaved by the blast. The ground was littered with fallen dead trunks, their roots stripped away. The dead trees bore traces of a continuous burn from above. Even the broken limbs of those trees still upright were charred at the break." The most astonishing statement from the scientist is the fact that every broken branch showed signs of fire, indicating that the burns were not those of a forest fire but the result of a sudden instantaneous scorching. A flash of intense heat seared and charred everything. As far as the eye could see stretched enormous dark patches of scorched, flattened forests. Trees uprooted and lay down at the same angle, their tops facing south or southeast. Moreover, they reported that "first, it seemed that the flying object had entered the Earth's atmosphere and become visible somewhere over Lake Baikal and then travelled from south-east to north-east as it plunged downwards, though there was some suggestion that it might have changed direction."

Interestingly, there were more than seven hundred eyewitness accounts. As a consequence, some Russian scientists believed that the object was a spaceship. Furthermore, Soviet academician Zolotov said that the object had been traveling "in cosmic terms extremely slowly before it exploded. Perhaps as little as one kilometer per second!" Some eyewitnesses suggested that the cylindrical object changed course. Also the irregular shape of the ground damage was like "an outspread eagle's wings." The point of explosion should have been more circular like an asteroid shape. Further, the Soviet scientists thought that the explosion took place in some kind of container. Moreover, if the object was a comet, it would be visible for more time, probably in many parts of the world. On the other hand, the object that exploded in the Tunguska forest it seems was sighted in the final phase of its trajectory before it exploded.

By the same token, in 1946, a scientist named Aleksandr Kazantsev postulated the theory that the explosion at the Tunguska forest was caused by a nuclear spaceship from another planet. He stated, "The explosion wave rushed downward and the trees directly below the point of the explosion remained standing, having lost only their crowns and branches. The wave burned the points of those breaks on the trees and with the permafrost, spitting underground waters, responding to the tremendous pressure of the blow, gushed up as those fountains seen by natives after the explosion. But where the explosion wave struck at an angle, trees were felled in a fanlike pattern. At the moment of the explosion, the temperature rose to tens of millions of degrees. Elements even those not involved in the explosion directly were vaporized and in part carried into the upper strata of the atmosphere where continuing their radioactive disintegration, that cause luminescent air. In part these fell to the ground as precipitation with radioactive effects."

As a matter of fact, a Tungus tribesman recalled, "My father went into the fallen taiga and saw a huge column of water flowing out of the ground. A few days later, he died in terrible pain as if he was on fire. But there was no trace of fire anywhere on his body." Likewise, numerous eyewitnesses who saw the object before it exploded in the sky reported that it was of considerable size and was a cylindrical shape. Also, the witnesses described the object in the form of a chimney. Some witness said, "This elongated flaming object glowed with a bluish-white radiance, brighter than the Sun and left a trail of multicolored smoke in the atmosphere." According to the Russian scientists, "In its descent over the Tunguska region, the object created a huge ballistic wave that was, exactly the same as the air wave of a missile." The velocity of this cylindrical missile-like object was first thought to be as much as 30 to 40 miles per second.

In order to account for the higher kinetic energy of the blast, however, the Russian geophysicist Zolotov made more accurate calculations of the speed of the object. He based his calculations from comparisons of the effect of the ballistic wave and blast force on trees in area of the explosion. He calculated that shortly before the explosion, "The velocity was probably no more than 1½ to 2 miles per second or about 700 miles per hour." Another Russian scientist, Professor Ziggel, points out that eyewitness saw the object and heard its deafening roar simultaneously. Interestingly, there are a couple of theories about what exploded in the Tunguska forest in Siberia; one is that it was a black hole or antimatter that exploded in the Tunguska. Although this theory doesn't account for the slow cylinder tube-shaped object described by seven hundred witnesses throughout the Siberian tundra.

Moreover, the object left a streaming trail, spaced explosions the oddly shaped pattern of leveled trees also the sudden growth of vegetation after the explosion. According to the Russian scientists, it is possible that the object was a spaceship from an extraterrestrial civilization. In 1961, Russian scientists showed the peculiar oval or elliptical shape of the Tungusca blast. A pattern that puzzled the Russian scientists was a trajectory approaching from east to southeast. A Russian scientist, E. L. Krinov, during the third Tunguska expedition in 1929, observed, "The area of uprooted forest has an oval form with the major axis situated in a direction from southeast toward northwest."

The oval shape surprised the Russian scientists because they expected a circular shape in accord with the crash of a meteorite or probably a comet. Certainly, the 1938 aerial photographs further verified the oval pattern at the site of the blast. Similarly, R. P. Florensky's expeditions in 1958, 1961, and 1962 determined from extensive ground and air survey and confirmed that the 1,200 square miles of forest leveled by the explosion and the scattering of cosmic dust from the blast had a definitive elliptical contour. Undoubtedly, in my opinion, only UFOs can cause this oval-elliptical form, of which there is plenty of evidence. Moreover, maps drawn by Florensky's team showed clearly that "the center of the destruction, the complete scorched area contained the dead but upright trees, lay in an off-center position in an explosive wave that fanned chiefly toward the south and north-east."

The Russian scientist wrote about the significance of the odd elliptical contour of the blast. He wrote, "It is very evident on the map of the region that the boundary of the area of complete leveling of the forest is irregular in outline. Also the epicenter of the explosion and the zone of trees left standing, occupy an eccentric position, in the region of the catastrophe. Obviously, this asymmetry cannot be explained by the effect of the ballistic wave, due to the flight of the body the zone of destruction is elongated in a direction that is not parallel to the trajectory but at a large angle to it." In fact, he characterizes the blast as directive because the effect of the explosion was "not the same in all directions."

Following two expeditions, one in 1959 and another in 1960, in which all the evidence about the Tunguska explosion was reexamined, the scientist A. V. Zolov came up with an explanation that Professor Zigel and the other experts found acceptable. He said, "The blast had an unusual oval shape, because the explosive material was encased in some type of container. The structure of the container, like the thick paper cylinder of a large firecracker caused two explosive charge, to fan out elliptically as it burst. The directivity of the explosion was due to the inhomogeneity of the container."

According to these scientists, the object that exploded at the Tunguska forest consisted of at least two parts: (1) A substance capable of nuclear explosion and (2) a non-explosive shell. Nonetheless, is there any evidence of a nonexplosive container? Indeed, some Russian scientists believed that at least partial proof was found by Kulik in his first expedition. Next, the Russian scientist A. Y. Manotskov made some calculations that agreed with Zolotov's findings that the object must have arrived at a velocity much slower, than that of a natural cosmic body. The object entry speed is comparable to the velocity of a high-altitude reconnaissance plane.

Another Russian scientist, a rocket specialist, examined this evidence and concluded with the other scientists that the object in its entry and velocity behaved like a supersonic craft. What flight path did this craft follow through the earth's atmosphere? Obviously, the principal clues to its trajectory are the reports and observations of eyewitnesses and the ballistic shock damage caused by the rapid compression of air ahead of the moving object.

The deafening thunder heard in June 1908 by hundreds of people throughout Central Siberia during the flight of the object was probably caused by its powerful ballistic waves. Similarly, the series of thunderclaps heard resulted from the massive blast waves. According to the Russian experts, the object created a strong ballistic wave during its flight that was exactly like the shock wave of a missile.

Furthermore, the Russian scientists studying the various investigations conducted since the 1920s to the 1960s have arrived at very shocking conclusions about the flight path of the object that exploded at the Tunguska forest. Three of the first researchers of the Tunguska explosion, Voznesensky, Suslov, and Astapovich, after collecting eyewitness reports and seismic data, stated that the object moved from south-southeast to north-northeast. According to Florensky's findings, "Both the general pattern of the toppled trees and the relationship between of fallen deadwood and the searing effects. As well as the distribution of cosmic dust indicated that the object came from east-southeast." Similarly, the scientist V. G. Konenkin said, "East-south-east to west-northwest was the most probable trajectory."

On the other hand, another scientist named Zolotov examined standing trees that bore traces of the ballistic and blast shock. He concluded that the air wave, which caused relatively minor damage compared to the explosion, had definitely come from the southwest. He added, "The object had been visible overhead as a 'fiery body' to villages near Kansk southwest of the blast, but it had also been seen in Kirensk and other towns lying to the southeast."

According to the scientists, dozens of reliable observations made both flights paths equally possible. Some scientist objected, "The same object could not have appeared almost simultaneously in two different locations hundreds of miles apart or could it."

Ultimately, the scientist solved the problem of the trajectory of the object with a shocking answer: "Both paths were accurate, the object had switched direction in its journey over Siberia."

Furthermore, the research by Florensky and Zolotev about the ballistic shock effect on the trees provides a strong basis for a reconstruction of an alteration in the object's line of flight. According to the scientist, the object "in the terminal phase of its decent the object appeared to have approached on an eastward course then changed course westward over the region before exploding. The ballistic wave evidence in fact indicates that some type of flight correction was performed in the atmosphere."

The Russian scientist Felix Zigel postulated the same theory. He was an aerodynamics professor at the Moscow Institute of Aviation and had been involved in the training of many Russian cosmonauts. His study of all the eyewitness and physical data convinced him and all the scientists researching the phenomenon that the object prior to exploding changed from an eastward to a westward direction over the Tunguska forest. Professor Zigel added, "Before the blast, the Tunguska object described in the atmosphere a tremendous arc of about 375 miles in extent in azimuth that it carried out a maneuver. No natural object is capable of such a feat!"

Consequently, the Russian scientists and rocket and aviation experts such as Manotzkov, Liapunov, and Kazantsev joined Zigel, Zolotov, and others in agreeing that the "cylindrical object causing an elliptically-shaped atomic blast in 1908 could only have been an artificial flying craft from some other planet. In addition to its maneuvers, near the earth's surface, the craft must have steered, as it approached from outer space, into a trajectory angle almost identical to the re-entry path used by modern spaceships." Obviously, the only candidates able to perform such feats are UFOs.

The Scientists' Findings in the Soil at the Tunguska Forest

The scientists found "tiny particles, little magnetite and silicate globules buried in the soil and embedded in the trees." According to the scientists, both are definitely extraterrestrial because the magnetite contains too much nickel, which is not found on this planet. Also the silicate had little bubbles of gas similar to that known from spectrum analysis of space objects. Moreover, the magnetite globules contain some very exotic rare earth elements including ytterbium. The Tunguska soil contained concentrations of spherical globules a few millimeters or less in size composed primarily of silicate and magnetite. A magnetized iron oxide, which resembled little droplets or bulbs, was sometimes linked together in clusters. Furthermore, the Russian scientist Krinov, who studied the soil, and A. A. Yavnel reported, "There are even instances where a magnetite globule was discovered in a completely transparent silicate globule. Those separate particles appeared to coalesce under tremendous heat, like the trinitite found at the Almogordo Atomic test site. The spheres were not of terrestrial origin."

Further, the "tiny brilliant spheres were found embedded like pellets in the earth and trees." Also, small amounts of cobalt, nickel, copper, and germanium.

Physical Evidence on UFOs
Spectrographic Analysis of Unknown Metals from a UFO Crash
Origin: Ubatuba, São Paulo, Brazil
Date: October 24, 1957

A sample from a UFO crash was a small piece of a silvery white metal, slightly oxidized on its surface and with a very low specific gravity. The spectrographic analysis of the metal revealed that it was made of absolute pure magnesium. Interestingly, no other metal or impurity was detected. Even a little trace of the commonly found metals were not present.

The conclusion reached by the scientists was that the quality of the magnesium was better than the purest magnesium refined on this planet (in 1957). Obviously, it represented something beyond the range of technological developments in earth science. So the scientist, based on this evidence, thought that the metallic pieces found on the beach near Ubatuba, São Paulo, were extraterrestrial in origin and part of a UFO.

ROBERT ITURRALDE

The Extinction of the Dinosaurs

> "Doom was coming out of the sky, in the form of an enormous comet or asteroid—we are still not sure which it was. Probably ten kilometers across travelling tens of kilometers a second, its energy of motion had the destructive capability of a hundred million Hydrogen bombs. If an asteroid, it was an inert, crater-scarred rock. dark and sinister, invisible until the last moment before it struck."

T. Rex and the Crater of Doom
Walter Alvarez—Geophysicist, Member of the Academy of Sciences
The Chelyabinsk Meteor

On February 15, 2013, in the Russian city of Cheliabinsk, a meteor crashed very close to the populated areas. Scientists thought that the meteor was about 60 feet in diameter and came undetected at roughly 42,000 mph and was almost 15 miles high when it blew apart. Fortunately, there were no deaths, and most of the 1,500 injured were from glass, as widows shattered when a shock wave hit the city 88 seconds later. According to scientists, if the meteor had detonated on the ground, it would have been a disaster. Also, scientists said that the meteor was an ordinary stony chondrite and not a rarer iron-nickel one, that it might have reached the ground. The Canadian scientific team calculated that the energy released by the meteor explosion was the equivalent of about 440 kilotons of TNT or about 30 times the power of the Hiroshima bomb.

The scientists' conclusion, with the help of data from a network of acoustic sensors, was set up to monitor compliance with the treaty to nuclear weapons' illegal testing. Actually, there are forty-five of the sensors in the world, detecting .50 called infrasound at frequencies below the human range of hearing. Furthermore, a spokesman for the preparatory commission for the comprehensive test ban treaty organization stated, "The explosion was detected by more than 20 of the sensors, including one in Antarctica, nearly 10,000 miles from

ESSAY ON THE THEORY OF THE EARTH: ELECTROMAGNETISM IN UFOS AND THE ORIGIN OF MASS EXTINCTIONS AND THE ICE AGES

Chelyabinsk. As a matter of fact, some of the detectors picked up the explosion more than once as the sound circled the planet several times. In addition, scientists with the information of how fast the meteor was traveling, they were able to calculate the mass, which was 11,000 metric tons." According to scientists researching this event, "It's all about the kinetic energy of that body which is related to its mass and velocity. The meteor has a certain amount of energy as it enters the atmosphere as it hits the air, it starts to decelerate quickly, and then the stress differential between the super-pressurized air in front and the less pressurized air behind, causes the rock to break apart violently." Furthermore, scientists said, "All that kinetic energy has to be released, only small fragments of the meteor reached the ground."

Next, I will present the list of the largest asteroid craters ever found on the planet. Nonetheless, not every impact results in a mass extinction.

Name	Location	Diameter (km)	Age (Millions of Years)
Vredefort	- South Africa	- 300 -	2023
Sudbury	- Ontario, Canada	- 200 -	1850
Chicxulub	- Yucatan, Mexico	- 180	-65(dinosaurs' extinction)
Manicouagan	Quebec, Canada	- 100 -	214
Popigal	- Siberia, Russia	- 100 -	35
Woodleigh	- Western Australia	90 ? -	?
Acraman	- South Australia	- 85-90 -	580
Chesapeake Bay	- Virginia, USA	- 85 -	35
Puchezh—Katunki	- Russia	- 80 -	175
Morokweng	- South Africa	- 70 -	145
Kara	- Russia	- 65 -	57
Beaverhead	- Montana, USA	60 -	>600
Tookoonooka	- Queensland, Australia	55 -	128
Siljan Ring	- Sweden	- 55 -	368
Kara-Kul	- Tajikistan	- 45 -	75
Montagnais	- Nova Scotia, Canada	- 45 -	50
Araguainha	- Brazil	- 40 -	247
Mjolnir	- Norway	- 40 -	144
Saint Martin	- Manitoba, Canada	- 40 -	220

As we see, of twenty asteroid craters, only one is considered as responsible for a mass extinction, the crater in Chicxulub in the Yucatan Peninsula in Mexico. Interestingly, some of the craters are bigger than the one assumed to have caused the dinosaur mass extinction.

The Extinction of the Dinosaurs

"Doom was coming out of the sky in the form of an enormous comet or asteroid. We are still not sure which it was." This is the beginning of the first chapter of the book *T. Rex and the Crater of Doom by Walter Alvarez*, a geophysicist, member of the Academy of Sciences.

The most important piece of evidence of an impacting asteroid is the molten material found in the Yucatan Peninsula in Mexico. The material appears in the form of spherules. The spherules were found in the boundary clay from samples from Spain. The crystal arrangement of the spherules suggested formation while molten; their composition and large size (too large for atmospheric transport) suggested that they were not volcanic. Furthermore, the scientists found that the spherules contained high concentrations of iridium. Shockingly, scientists who believed the asteroid impact theory were not ready to accept the findings. Likewise, some scientists thought that perhaps it was a cometary collision. The impacting body would have been very large to have left such a large amount of iridium. The scientist stated that only three craters 100 km or more in diameter are known.

Two are pre-Cambrian and the too young, also two-thirds of the Earth is covered with oceans. So the asteroid probably had plunged into the ocean, and with the seafloor spreading, perhaps it had disappeared into the Earth's mantle. The Cretaceous mass extinction, it seemed, had occurred throughout the land and sea in the world. Moreover, other scientists stated that the uneven levels of extinction appeared to contradict a sudden mass extinction. The scientist Michael R. Rampino of NASA's Goddard Institute for Space Studies had other theory to explain the abnormal amounts of iridium. He said that perhaps a change in seawater chemistry altered sedimentation or caused a hiatus in sedimentation of other elements, thus causing an abnormal concentration of iridium. Although, changes in seawater chemistry would not explain the iridium in the clay formations of New Mexico.

The impact might be a fact, but how did it trigger the mass extinction of the dinosaurs and thousands of species? Moreover, scientists argue that the fossil record does not show that a single event caused the mass extinction. For instance, they said that there was no obvious pattern from the lists of those plants and animals that died off and those that survived. For example, marine plankton species were exterminated. On the other hand, freshwater organisms survived. Land plants in North America were devastated, but no tropical plants. Pterosaurs died but not birds; most of the marsupials died off but not the mammals. The placentals and multituberculates and ammonites went extinct, but not their relatives, the squid. Furthermore, dinosaurs vanished but not their fellow archosaurs, crocodiles. Also, small dinosaurs were gone; interestingly, reptiles of big size survived. The only mammals that survived were the size of shrews. The extinctions were widespread and worldwide. Interestingly the extinctions were highly selective; that is the reason that only one asteroid cannot have triggered a selective mass extinction.

Interestingly, scientists studying the Spanish KT boundary samples from Caravaca discovered some sand, grain-sized, rounded white objects of peculiar composition, which they called spherules. The spherules had a clue to the location of the crater left by the asteroid in land or ocean. In fact, the spherules haven't been understood totally by the scientific community. Conversely, the scientists using the geologists' method for studying rocks and minerals cut the spherules in half, glued them to a piece of glass, and ground them so thin that they became transparent. Surely, after studying these thin sections with a microscope, they found that the internal crystal structure was feathery.

A very strange shape for a mineral grain! the scientists thought. Furthermore, when analyzed chemically with the electron microprobe, the scientists found that the feathery crystals were made of the mineral Sanidine, a kind of potassium. A very strange mineral to find in a sedimentary rock. Likewise, scientists were able to use oxygen-isotope ratios and found that the evidence in the spherules was not original. They stated, "It came neither from the impactor nor from the target, but instead was a replacement mineral that had grown later. The original minerals of the spherules had been something different." By the same token, scientists sampled many KT boundaries in Italy, and also they found spherules. After the scientists studied the spherules under the microscope, they found more of the "unusual crystal textures in some plalike the branches on a snowflake." The scientists agree that these were the textures "of olivine, pyroxene and calcium, rich feldspar crystallized from molten rock at an unusual, intermediate rate of cooling." The scientists stated, "Neither the slow cooling that produces full crystals nor the quick cooling that produce glass." Furthermore, the scientists reported that the original minerals in the KT spherules had been olivine, pyroxene, and calcium-rich feldspar, similar to what the scientists found in a KT site in the sediments of the floor in the Pacific Ocean.

So the scientists concluded that it was an impact on the oceanic crust. The bedding in the area of the impact was found "packed with spherules about one millimeter across with bubbles in it. The impactor struck and the crater was formed, on the north coast of what today is the Yucatan Peninsula."

Undoubtedly, the similarities between the Tunguska explosion and the asteroid impact during the dinosaurs' extinction are in the findings after the impact. In the Tunguska soil were found concentrations of spherical globules a few millimeters or less in size, mostly composed of silicate and magnetite. A magnetized iron oxide. The magnetic globules resembled little droplets or bulbs and were sometimes linked together in clusters. The experts stated, "There are even instances where a magnetite globule was discovered in a completely transparent silicate globule." These separate particles appeared to coalesce under tremendous heat, like the trinitite found at the Alamogordo Atomic Test site.

In the Tunguska site, scientists found thousands of tiny brilliant spheres; many fused together were found embedded like pellets in the earth and trees. The spheres revealed small amounts of cobalt,

ESSAY ON THE THEORY OF THE EARTH: ELECTROMAGNETISM IN UFOS AND THE ORIGIN OF MASS EXTINCTIONS AND THE ICE AGES

nickel, copper, and germanium. Furthermore, as long ago as 1872, the Challenger oceanographic expedition, which sailed over 68,000 thousand nautical miles in circumnavigating the globe, recorded the presence on the world's ocean floors of thousands of small black nickel-rich magnetic spherules having diameters in the order of 1050 microns. They appeared to be especially abundant in the deepest basins of the oceans. The existence of these tiny objects was, a couple of times, confirmed in the late 1940s by the Swedish marine exploration vessel Albatross. Some scientists thought that they probably were of meteoric origin. Interestingly, the scientific research team didn't find signs of craters that could point to an asteroid or meteoric impact.

I doubt that only one asteroid was the cause of the mass extinction of the dinosaurs. The dinosaurs were the most successful animal species that ever walked this planet. As a matter of fact, the dinosaurs lived for more than 140 million years, and they were adapted to every climate and continent, including Antarctica. So I think one asteroid cannot trigger the extinction of the dinosaurs. Certainly, it was not the fate or destiny nor was in the DNA of the dinosaurs to go extinct. Nor had nature planned the extinction of the dinosaurs because there is nothing within Earth that can cause a mass extinction. Moreover, nature had no plans to allow the mammal species to evolve and become the dominant species. All mass extinctions, at least on this planet, are caused by an external force, alien in nature. In our case, the only candidate that I know is the intelligence behind the UFO phenomenon, which some people call God. The only species capable of self-destruction and as a consequence of mass extinction is humankind. Although, it is not written in our DNA that we have to become extinct, because if we not in our DNA written that we have to become extinct, because if we transcend those pitiful desires of power, greed, and religious fanaticism, nothing would be in the way for humankind to exist as long as the dinosaurs. The intelligence behind the UFO phenomenon had planned all the mass extinctions, which allowed evolution to work and intelligence to appear with the arrival of the mammal species.

Interestingly, in a museum in Salzburg, Austria, there is a perfect little cube of meteoric nickel-iron, which is a form of steel that is about two inches square, though the opposing sides are convex or domed. The object weighs 28 ounces and has a specific gravity of 7.75. A precisely turned deep groove or incision completely encircles it. Undoubtedly, the object has been examined endlessly by scientists, and their conclusions are shocking. The researchers stated that because the components are nickel and iron, the object definitely was constructed by machine manufacture and comes from a block of a meteorite. Although the most amazing thing is that the object was found in a block of coal taken from a Miocene stratum 65 million years ago, the same time of the extinction of the dinosaurs! Some will choose an easy answer—that is just a hoax. My theory is that the intelligence behind the UFO phenomenon is responsible and it was left there with the purpose of making us aware of their presence on the planet. Also, I don't doubt the possibility that they redirected the asteroid or asteroids to crash on the planet and caused he mass extinction of the dinosaurs.

Impossible? I answer that is not impossible right now; NASA is making plans to capture an asteroid and, if possible, to explore for minerals. NASA wants to launch an unmanned spacecraft in 2018 that would capture a small asteroid. The idea is to haul it closer to Earth then send astronauts up to examine it. First they want to build a robotic spacecraft with a cone-shaped structure that will envelop the asteroid. After wrapping the asteroid in a bag (no lassos are actually involved), a manned spacecraft would meet up with the captured asteroid, allowing astronauts to study it. NASA called their project the Asteroid Lasso Plan. So it is not impossible for the UFO intelligence to have perfected this technology.

CHAPTER 5
The Strange Death of the Megafauna and Flora during the last Pleistocene Mass Extinction

The pole of the earth moves in a circle round the pole of the ecliptic, and its axis is more inclined to the plane of the ecliptic; but these two motions, the causes of which are now ascertained, are confined within certain bounds, and are much too limited for the production of those effects which we have stated. Besides, as these motions are exceedingly slow, they are altogether inadequate to account for catastrophes which must have been sudden. The same reasoning applies to all other slow motions which have been conceived as causes of the revolutions on the surface of our Earth, chosen doubtless in the hope that their existence could not be denied, as it might always be asserted that their extreme slowness rendered them imperceptible. But it is no importance whether these assumed slow motions be true or false, for they explain nothing, since no cause acting slowly could possibly have produced sudden effects. Admitting that there was a gradual diminution of the waters; that the sea might take away solid matters from one place and carry them to another; that the temperature of the globe may have diminished or increased; none of these causes could have overthrown our strata; enclosed great quadrupeds with their flesh and skin in ice; laid dry—sea shells in as perfect preservation as if just drawn up alive from the bottom of the ocean; or utterly destroyed many species, and even entire genera, of testaceous animals.

—Georges Cuvier

The strange death of the mammoth and other animal species during the last Pleistocene mass extinction is one of the great mysteries in natural history. I will start with the extinction of the mammoth, which is the most perplexing puzzle to solve. The mammoth, up until 11,000 to 12,000 years ago at the end of the last Ice Age, was the most successful quadruped. He already had been through more than half dozen Ice Ages. So my question is, why did the entire species suddenly disappear 11,000 to 12,000 years ago? Moreover, more than fifty million animals disappeared in the United States alone and millions more in Canada, Europe, Africa, Australia, and South America.

The woolly mammoth is known to science as Mammuthus primigenius, and in the last two hundred years, very well-preserved carcasses have been uncovered. According to scientists, the mammoth appeared in Europe and Asia and is of the elephant family. The mammoth thrived in North America, Canada, and Alaska before becoming extinct. Furthermore, the wooly mammoth had been found in Wyoming, Lake Michigan, and Siberia. In fact, in Siberia, millions of mammoth remains have been found, and the ivory of the giant tusks up to 15 feet long was in perfect condition to make ornaments. In August 1799, an almost complete corpse of a mammoth was found in the delta of the Lena River. Scientists began to ask, why have the creatures died out? It was shocking that their corpses were in such great condition that their meat was eaten by dogs.

Likewise, in August 1900, a group of hunters found a mammoth carcass on the banks of the Berezovka River in the province of Yakutsk in Northern Siberia. Moreover, carbon dating analysis indicated that the mammoth died between 39,000 and 47,500 years ago. Shockingly, in the mouth of the mammoth were found buttercups and grass, which for some reason it was unable to swallow. Obviously, this event indicated that the animal died suddenly.

More interestingly, some of the plants found in the stomach of the mammoth still grows in the area. Furthermore, an analysis of the mammoth's contents revealed that contained grasses, mosses, and lichens of various kinds, also the green branches of tundra trees as fir and pine. Shockingly, the presence of seeds showed that the mammoth had been frozen during the second half of July or August.

Scientists believed that the mammoth had been browsing when it stepped onto thin ice and plunged into the shallow ravine, breaking its legs and pelvis. As a consequence, frozen slush and snow from both sides of the bank had suffocated. A late great scholar stated, "The evidence shows that the animal suffered a very severe fall, severe enough to break his pelvis and leg." Also, large amounts of blood were found under his body; if the animal had fallen in the river, the blood would have washed away.

Another fact about the Berezovka mammoth is that his penis was found to be erect. This fact indicates that the animal was not instantly killed by his fall but that he suddenly froze to death. Moreover, the frozen ground or permafrost might have extended down thousands of feet under the surface. According to scientists, the erection of the penis could have resulted from the poor's animal terror and pain.

Furthermore, the permafrost or frozen ground is covered with a layer of soil called muck. The muck sometimes thaws in the summer; it is composed of mud, silt, and black organic matter bound together with ice. Many scientists assumed that these animals fell into the ice, although some scientists suggested that it is impossible. After all, there are no—and never were—glaciers in Siberia where the mammoths were found, although in the upper slopes of few mountains there is some ice. Interestingly, the animals were never found in mountains but usually on the plains, a little above sea level. Moreover, all these animals never have been found in ice, only in the muck. In addition, the animal remains are not found in deltas or stuaries but in the plateaus, all over the tundra between the river valleys. Interestingly, there is no evidence of glaciers or an Ice Age in Siberia, so how did these animals freeze to death? The Berezovka mammoth was discovered in a squatting position, raised up on one foreleg. The head had been mostly eaten to the bone by wolves, but much of the body was in good condition although the lips and the tongue were preserved with his last meal, which it had no time to swallow. Scientists found buttercups in his mouth that even today bloom in the summer months. Obviously, scientists agree that freezing is not a simple process as some might think. Specialists of the food industry stated that to preserve meat properly, the following steps must be taken: "Meat must be frozen very rapidly. If it is frozen slowly large crystals form in the liquids in its cells. These crystals burst the cells and the meat begins to deteriorate at 40 Fahrenheit it takes 20 minutes to quick freeze a dead turkey, 30 minutes to do a side of beef," although these are only small amounts of meat, nothing like a wooly mammoth with heavy fur. Interestingly, the center of the body of the mammoth remains warm, and the freezing of the flesh must be slow enough for large crystals of ice to form within cells, because rupturing them would cause the flesh to deteriorate. Shockingly, neither event took place with the thousands of mammoths found after eleven thousand years of being frozen. Actually, the meat of a mammoth was sampled by British scientist members of the Royal Society without any ill results. Furthermore, almost everywhere, mammoths were found in the plateaus between the river valleys. Also, it was pointed out that the whole of northern Asia, Alaska, and Western Canada could never have been one vast delta. Conversely, the rivers never could have deposited muck uphill. As a matter of fact, thousands of these animals were found in perfect condition. The animals were fresh, whole, and without any damage that suggested being transported by water. As a matter of fact, many animals have been found standing or keeling upright. Obviously, a mammoth falling into a river is not going to be carried away in an upright position and deposited hundreds of miles away. Actually, mammoths and elephants are very good swimmers. Also because of the great contents of fat in their stomachs, they can float for long time.

However, these standing mammoths with their fur coats were in perfect condition.

Likewise, there was a mud theory that claims that are certain kinds of clays found on the tundra only. Few inches of the clay are sticky enough to hold a man by his feet. So probably it can hold a mammoth until it freezes to death. Although, such a substance never has been found holding or next to a mammoth. However, the existence of this type of clay had to be unfrozen at the time, and that the temperature of the air was well above freezing. Another theory is that after the mammoths got stuck in the clay, a giant blizzard blew up both the mammoth and the clay. Another interesting thing about the Beresovka mammoth is that it was found "like squatting at the back end, but was raised on one foreleg in front, with the other held forward as if about to salute." The mammoth was perfectly frozen for their cells not to burst and survived for more than ten thousand years.

Some scientists stated, "The problem is extract all the heat from the whole beast, but this can only be done from the outside and by working inward, unless we have tremendous cold outside, the center of the animal and notably its stomach, will remain comparatively warm for some time." Interestingly temperatures of lower than minus 100 degrees Fahrenheit in Antarctica and in other parts of the Earth less than zero below have been recorded. Still, a very large number of animals survived, like sled dogs burrowing into the snow to sleep in Antarctica. Moreover, it is known that human beings well clothed have endured temperatures of minus 100 degrees for up to a half hour. As a matter of fact, it takes a very strong cold to kill a warm-blooded animal.

Furthermore, according to some biologists, mammoths, though covered with a thick underwool and a long overcoat, "were not specially designed for Arctic conditions." They didn't live all year round on the Arctic tundra. For instance, their close relative, the Indian elephant, who is about the same size, needs several hundred pounds of food daily just to survive for more than six months. Certainly, there is nothing for the mammoths to eat in the tundra. Some scientist said that buttercups "will not grow even at 40 degrees and they cannot flower in the absence of sunlight."

Furthermore, a detailed analysis of the contents of the Beresovka mammoth showed a number of plants that still grows in the arctic. However, the plants are more typical of southern Siberia today. So probably, the mammoths made annual migrations north for the summer. Or that part of Siberia where the mammoths were found was in warmer latitudes. Most astonishing is that ivory which is easily destroyed by exposure to the weather has been preserved in almost perfect condition for trade since the 1700s. Millions of ivories have been found in the New Siberian islands.

More shocking is the conclusion of the French Dermatologist H. Neuville, who published a report about the mammoth in 1919. He stated that it is not true that the mammoth was adapted to a very cold climate. He said that the mammoth's thick skin, hairy coat, and the deposit of fat under the skin were not necessarily an adaptation to cold climate. He performed a comparative microscopic study of sections of the skin of a mammoth and an Indian elephant and discovered that they were identical in thickness and in structure. The skins were not only similar but exactly the same! Also he stated that the lack of oil glands in the skin of both animals made their hair less resistant to cold and damp than the hair of the average mammal! So the hair of the mammoth showed a negative adaptation to cold. So he stated that the common sheep is better adapted to arctic climate than the mammoth! He said that the mammoth lacks sebaceous glands and that the oil produced by these glands is important for the protection of the animal against the cold. Obviously, oil in the hair blocks the penetration of cold. The hair of the mammoth without oil would offer poor protection against the arctic cold. Many biologists have said that many animals in the jungles of Africa and the Amazonas such as lions, tigers, etc., have fur, which by itself does not mean an adaptation to cold. Moreover, Neuville stated that fur without oil is "a feature of adaptation to warmth, not cold!"

Furthermore, there are more questions about the sudden death of the mammoth; for instance, how is possible that the body of the mammoth was deep frozen? How did it get into a solid mass of permafrost, also deep frozen, without destroying his body? Or did the outside of the body rot due to warming up and refreezing? Georges Cuvier wrote, "Everything thus seems to indicate that the event [cause] that buried them was one of the most recent that has contributed to changing the surface of the globe. This was nonetheless a physical and general event. The bones of fossil elephants are too numerous and found in too many deserted or even uninhabitable regions. For us to be able to suspect that they were taken by human beings. The beds that contain the bones and overlie them show that this event was aqueous, or that it was water that covered them up. And that in many localities this water was almost the same as that of the sea today. Since it sustained roughly similar organisms. But it was not this water that transported them to where they are."

In addition, the bones of the mammoth are found in almost the whole world. He proceeded, "An eruption of the sea that would have carried them only from the areas where the Indian Elephant now lives. Could not have spread them so far nor dispersed them so uniformly. Besides, the inundation that buried them did not rise over the major mountain chains. For the beds that it deposited and that contain the bones are found only in low lying plains. Thus one cannot see how the carcasses of Elephants, Mammoths, Mastodons etc. Could have been transported to the north [Siberia] over the mountains of Tibet and the chains of the Altai and the Urals."

Moreover, another question is if the Beresovka mammoth died and was frozen 39,000 to 47, 500 years ago in the same place where it was found or if it was transported already frozen. How did a frozen mammoth get stuck above ground into a stratum of frozen muck that is rock-hard? Now if was transported by water and already frozen, why are there no glaciers in Siberia? More questions—how did hundreds of thousands of mammoths and megafauna end up in the Siberian islands? Now the elephants are very good swimmers; they can also float, but there are two hundred miles of ocean. So how were the thousands of mammoths able to reach the islands, because even large islands are easy to miss? Further, the greatest of the Siberian islands is about 150 miles long and about half as wide. Explorers of the Siberian islands found mammoths and other megafauna also in the interior of the islands. How is possible for the Lena River to inundate the New Siberian islands, two hundred miles at sea? Scientists point out that all the rivers of Europe and Asia put together cannot, at full flood capacity, raise the ocean level two hundred miles off the coast by little of a few inches.

By the same token, let's assume that they were washed to the coastlines and not into the interior. The question, is how were hundreds of thousands of mammoths and megafauna placed above high water level? Undoubtedly, storms can wash up and wash away and cause erosion; that is why nothing accumulates along the coasts.

Furthermore, why after so many thousands of years and warmed summers, which usually it will melt snow and ice fail to melt where the Mammoth stood and let it rot? Also, how did the mammoth get stuck in ice and muck? So there probably never was ice or a tide of muck that came after freezing, although the mammoth never would have been overtaken by muck because muck also would have been completely frozen. Shockingly, the permafrost where the mammoth was found was 11,000 to 12,000 years old! How can we reconcile the age of the mammoth? According to a Carbon-14 test, it was 39,000 to 47,500 years and the age of the permafrost? My theory is that the intelligence behind the UFO phenomenon to cause mass extinctions used a technology that changed the structure of time and space. As I mentioned before, all over the world are geological ages in the stratum missing or in the wrong order. Definitely this is not the logical way nature should work.

The technology that the UFO phenomenon probably used is already known by scientists. For instance, scientists are trying to determine whether faster-than-light travel is possible based on the findings of a 1994 Mexican physicist, Miguel Alcubierre. He theorized that faster-than-light speeds were possible in a way that did not contradict Einstein. His theory postulated the expansion and contraction of space. So the spaceship would manipulate space and time and travel near the speed of light. In addition, such a machine distorted time and space, which is exactly what we see with the mammoth age and the permafrost where it was found and the different age in stratum around the world.

Also, it is possible that time is an illusion and that we live in an eternal present and that the divisions of time are human conceptions. Einstein showed that the line between past and future varies with the observer. According to scientists, there have been sixteen or seventeen cold cycles during the last two million years, and the mammoth survived all of them. So why did they become extinct this last Ice Age? Moreover, hundreds of mastodons and mammoths have been found in the strangest positions against any common sense or natural works of nature. For example, nearly all the skeletons found in the deposits in the valley of the great Osage River were in a vertical position. Similarly, since colonial times, many mastodons have been found standing erect just below the surface of swamps or bogs. Likewise, the Newburgh mastodon was found, according to scientists, in this manner "The anterior extremities were extended under and in front of the head. The posterior extremities were extended forward under the body."

Also, in another case, a mastodon was discovered in Monmouth County, New Jersey, in 1823. It was described as follows: "Its vertebral column with all its joints and ribs attached to them in their natural position, lay about 8 or 10 inches below the surface. The scapula rested upon the heads of the humeri, and these in a vertical position, upon the bones of the forearm as in life. The forearm was still buried. It inclined a little backwards, and the foot which it was immediately below, it was placed in advance of the other, as it would be if the animal had been walking!" The four feet rested on the frozen ground. A scientist said, "Its position was vertical, the feet resting on a stratum of sand and gravel, and the head to the west-south-west."

Similarly, other mastodons and mammoths have been found lying in the strangest positions. Furthermore, a mastodon and his calf were found in a standing position in a very narrow space. How did these large animals get embedded in a narrow space? The scientists that found them thought that they probably were mired attempting to reach a spring. Although the basin was small and shallow with a bottom that was solid, solid paved with rolled stones. For the scientists, it was very difficult to think how, once the mammoth got stuck in thick mud, even a powerful animal like a mastodon could turn itself upside down and become buried on its back. Obviously, it might fall to one side and remain there, but not upside down! Likewise, the solid basin where the mammoths were found was very common.

Also, according to scientists, it seems that the animals may have sensed the catastrophe because of the strange southwest to northeast position of many of their remains. Certainly, the catastrophe was so sudden that there was not even time for these animals to be knocked over by it. A scientist stated, "The cause of the disappearance of the Mastodon and Mammoth seems to be mysterious, we are disposed to believe that an animal of so large a size, of so great strength and such extensive distribution must have required some great and general catastrophe to overwhelm and annihilate it! So geologically recent was their demise." The remains that are found in Alaska and Siberia are so fresh that they are indistinguishable from plants and animals that have died only weeks or months. Obviously, the form of death of all these giant animals is completely unnatural because the floods, earthquakes, or hurricanes produced these strange forms of death. I believed that mastodons and mammoths and even water and muck were teleported just like the erratic boulders—that is the reason they were found in so many strange positions. We know that natural catastrophes never show intentionality, and these findings show intelligence. Similar with the Beresovka mammoth that was frozen perfectly to last thousands of years.

Nature cannot produce such effects because it doesn't have consciousness. Moreover, the idea that the mammoth and megafauna extinction was caused by Stone Age hunters is a myth. Obviously, there is no evidence that hunters caused the last mass extinction. Probably the best example is that the American bison never got extinguished, even with the colonization of the American continent.

ESSAY ON THE THEORY OF THE EARTH: ELECTROMAGNETISM IN UFOS AND THE ORIGIN OF MASS EXTINCTIONS AND THE ICE AGES

The Near Extinction of the American Bison

According to anthropological studies, there were probably four to eight million people living in the United States in AD 1000. For the American Indian, the bison represented everything they needed for daily life, from meat to medicines, sacred objects, robes, skins, war clubs, knives, fuel, and toys. The women carved horns into cups and from the hooves into glue. Further, the bones were turned into sled runners or hoes; also, they braided the hair into lariats or stuffed pillows. Likewise, they made thread from sinews and containers from the stomach and bladder. In addition, they used the bison's shaggy tail in vealing ceremonies or as a fly switch. So even with the use of every part of the bison's body, the Native Americans didn't cause the extinction of the bison. The American bison, also known as the American buffalo, is a North American species of bison that once populated the American landscape. A bison has a life expectancy of fifteen years in the wild and twenty-five years in captivity.

Interestingly, the bison was never domesticated by the Native American. The American bison is a newcomer to the United States that originated in Eurasia and migrated through the Bering Strait perhaps ten thousand years ago. In fact, the American bison replaced the steppe bison (Bison priscus), a much larger bison. Interestingly, a belief is that the long-horned bison became extinct due to climate change and overhunting. The fact is that either theory is not truth. The Native American supplemented his food with fish and by killing and eating every size and kind of animal. The Indian tribes were split in different hunting bands, each confining to a determined territory. by the same token, the members of a band of one hundred Indians would require each a minimum of four pounds of meat a day. So the top five to ten hunters of the band would need to kill an average of four deer or one elk each day, or three or four moose or two buffalo per week. The Native Americans chase the bisons and wild animals on foot because the horse was reintroduced by the European pioneers. As a matter of fact, until the reintroduction of the horse by the Spaniards, all traveling by the Native American was done on foot. Moreover, even after the arrival of the horse, it was not common for every member of the tribe to own a horse. In most tribes, the members took turns riding and walking. Similarly, wheeled vehicles were unknown until they were brought to the Americas by the Europeans.

Obviously, pulling, carrying, and lifting were done through human muscle power. The Indians used arrowheads, spear points, axes, and lances or spears. A good archer could put an arrow into a moving target at fifty yards. As a matter of fact, a witness reported once during a competition an Indian drew fire so fast that he put eight arrows in the air before the first one reached the ground. Also, the Indians used the method of using fire to scare the buffalo to run to the edge of precipices. As matter of fact, as early as 1540, horses had been introduced across the Rio Grande into the southwest by the viceroy of New Spain, Vazquez Coronado.

Although by the 1650s the plain Indians owned a handful of horses, twenty years later, they had double their numbers. After, in 1581, 1582, 1590, and 1591, more horses were reintroduced. Interestingly, the Indians of the period 1600-1680 ate the horses instead of riding them. Probably by the 1680s the Indians began to ride the horse. Also, the French and English brought horses to the continent. So between 1650 and 1800, the American Indian used and rode horses in their hunting of the bison. He didn't have to walk or run to chase the bison. Next, after the introduction of the horse came the introduction of firearms, both gifts of the white man. Furthermore, before the arrival of the buffalo hunters in the 1870s, there were sixty to eighty million bison in the United States. The settlers hunted the bison for them from which dozens of accessories were made although the meat was left to rot. After the animal rotted, their bones were collected and shipped east in large quantities. Furthermore, the US army actively endorsed the wholesale slaughter of bison. Conversely, by the 1830s, the Comanche, their allies, and buffalo hunters were killing about 280,000 thousand bisons a year. Obviously, firearms and horses and a need for a growing market for bison robes and bison meat increased the hunt of bison by native Indians and the white man. The main reason for a near extinction of the bison was commercial market hunting just like man-made extinction of the passenger pigeon. Moreover, bison skins were used for clothing such as robes, rugs, and most important, for industrial machine belts. As a matter of fact, one professional market hunter killed over twenty thousand bison. After all, a good hide could bring three dollars in Dodge City, Kansas, and a very good one for a heavy winter coat could bring fifty dollars. Moreover, for a decade from 1873 on, there were over a thousand commercial-hide markets. The slaughter by these commercial enterprises surpassed what the Indian hunters killed just for daily consumption. These commercial enterprises killed from two thousand to one hundred thousand animals per day depending on the season.

In fact, it was said that the rifles, big .50s, were fired so much that the market hunters needed at least two rifles to let the barrels cool off. However, in 1874, President Ulysses S. Grant pocket vetoed a federal bill to protect the few thousand bison herds, and in 1875, General Philip Sheridan pleaded to a joint session of congress to slaughter the bison to deprive the Indians of their source of food. Surely, by 1884, the American bison was near extinction. So to end this segment, the reason for the near extinction of the American bison was the introduction of the horse, firearms, and the white man. Otherwise, the Native Americans never would have been able to hunt the bison to near extinction without the white man's help. Remember, before the reintroduction of the horse, the Indians hunted the bison on foot. So for the scientific community to say that prehistoric man hunted the mammoth and the rest of the megafauna to extinction and even small rodents is ingenuous and foolish. The overhunting theory was just too ingenuous to be truth; no event climate change could have the last Pleistocene mass extinction. A worldwide mass extinction in every continent and island. Scientists have found evidence of mass extinction all over the world and in places impossible for normal forces of nature to operate.

ROBERT ITURRALDE

Slaughter or Mass Extinction of Incrustations

It is also possible that the animals inhabiting shells may leave their stony coverings when they die in some particular places; and that these cemented together by slime of greater or less consistence, or by some other means, may form extensive banks of shells, but, we have no evidence that the sea has now the power of agglutinating these shells by such a compact paste, or indurated cement as that found in marbles and calcareous sand-stones or even in the coarse limestone strata in which shells are found enveloped. Still less do we now find the sea making any depositions at all of the more solid and silicious strata.

—Georges Cuvier (1769-1832)

For example, rivers bring down with them all kinds of earth serviceable for the growth of plants, which sometimes is deposited inland, often also at their mouths. The tide brings this mud to many coasts over the land or deposits it on the shore; and so, more especially if men give their aid so that the ebb shall not carry it back again, the fruit-bearing land increases in area, and the vegetable kingdom gains the place which formerly was the habitation of fish and shells. Now the question is whether this is to be judged a purpose of Nature because it contains profit for men.

Thus we cannot take for natural purposes rivers because they promote intercourse among inland peoples, mountains because they contain the sources of rivers and for their maintenance in rainless seasons have a store of snow, or the slope of the land which carries away the water and leaves the country dry because, although this shape of the earth's surface be very necessary for the origin and maintenance of the vegetable and animal kingdom, it has nothing in itself for the possibility of which we are forced, to assume a causality according to purposes.

Without men the whole of creation would be a mere waste, in vain, and without a final purpose.

—Immanuel Kant (1724-1804)

As the great paleontologist George Cuvier stated, the sea "has no power of agglutinating the shells into a compact paste." In other words, water or liquids in general have no material properties to hammer seashells or animal parts into crevices or little holes. Furthermore, the great philosopher Immanuel Kant stated that the natural world in itself and without mankind has no purpose. The incrustations of animals and seashells that scientists have found in crevices and fissures all over the world could not have been a effect of water or floods. The incrustations are not natural formations because there is intentionality in their formation, because if water has no material properties, then what created this incrustations?

Mediterranean Slaughterhouse

A great scholar, Dr. Hugh Falconer, after exploring the Sicilian cave of Moccagnore stated, "The cavern had been filled right to the roof, the uppermost layer consisting of a concrete of shells, bone splinters, with burnt clay, flint chips, bits of charcoal and hyaena coprolites, which was cemented to the roof by stalagmitic infiltration of contemporaneous origin occurred to the conditions previously existing, emptying out the whole of the loose, incoherent contents and leaving only portions agglutinated to the roof."

Moreover, in the hills of Palermo, Sicily, thousands of hippopotamus bones have been found. Experts stated that twenty tons of these bones were shipped from around the cave of Sanciro near Palermo. Within the first six months of exploiting them, they were so fresh that they were sent to Marseilles to furnish animal charcoal for use in sugar factories. Interestingly, the bones were of animals of all ages, even a fetus, and they didn't show traces of weathering or exposure. The scientists researching this site reached the conclusion that "the freshness of these bones and the presence of very young individuals show that some sudden and geologically recent cataclysm was responsible for their accumulation." Similarly, another scientist, Dr. Prestwich, stated, "Impossible to account for the phenomena by any agency of which our time has offered us experience. The agency whatever it was must have acted with sufficient violence to smash the bones."

Likewise, at the Moccagnore cave, scientists found a great variety of animals. For instance, they found hyenas, lions, bears and a tusked elephant, a dwarf form of a hippopotamus. The animals were found in caves and rock fissures often too small even for one beast. Also most of these animals did not live where they were found. In fact, it seems that they were transported from very distant regions. Moreover, remains of turtles and tortoise were found—fissures and whole carcasses, as a matter of fact. Because of the freshness of the bones, it seems that many animals have been entombed in the flesh and in dismembered condition. Shockingly, the individuals represented are of every age, from infant to adult. Furthermore, in a little fissure scientists found "among enormous quantities of fossilized shells, lay the two detached lower jaw bones of possibly the smallest of the two pygmy elephants, and under them portions of the spinal column with ribs."

Now the question is, is water in the form of a flood the agent responsible for this phenomenon? Then how did the flood get the power to agglutinate immense animals in fissures? So why do floods now have no power to agglutinate enormous carcasses? Actually, water, being transparent, has no material properties like a hammer to nail objects. Furthermore, scientists agreed "that there must have been several of the elephants introduced in the flesh!" Likewise, these kinds of fossils have been found in Eurasia and North and South America, and scientists agreed that "these animals were not dismembered by natural predators, but by something acting just before or during their burial on quite literally a hemispheric scale." Obviously, a flood can't dismember enormous animals and congregate them in a narrow creek. Also, in British sites, they found "young and old animals of many incompatible species entombed together, no single individual being preserved entire and the majority in an exceedingly fragmented condition, many bones have also been found jammed into the furthest recesses of the caves."

Of the intermix of different fauna and flora from different regions and latitudes. Further, in a cave in Bristol, England, scientists found carcasses of northern and southern animals. The northern animals were bison, reindeer, and woolly rhinoceros, and the southern species by the tusked elephant, the round-nose rhinoceros, and the hippopotamus. Similarly, at Nice, investigators found in a cave fissure deposits containing land shells, snakes, tortoises, and the bones of the woolly mammoths and wooly rhinoceroses. Likewise, in caves between Monaco and the Italian border, investigators found bones of lions, rhinoceroses, hyenas, macaco monkeys, and elephants. Interestingly, whales and other marine animals were found in the fissures and caves.

Similarly, a great profusion of shells and animal bones are found in breccias at Verona, Vicenza, and other regions in the south of Italy. Additionally, in caves and fissures near Beirut, Lebanon, shells, bones of bison, and other northern animals are found in large quantities. Similarly, remains of late Pleistocene elephants, the woolly rhinoceros and buffalos, and the Indian water buffalo mixed with southern animals have been found in a cave in Ksarakil, Syria. Scientists asked the question, "What were these northern species doing so far south?" Scientists can't explain this phenomenon.

Furthermore, animals of typical northern origin like bison, bear, and large carnivores have been found in Israel. Similarly, in the Altai Mountains of Central Asia, scientists have found mastodons, giant camels, gazelles, saiga antelopes, giant beavers and ostriches in fissures and caverns closely packed. Interestingly, sometimes the "bones were unabraded and looked amazingly fresh!" investigators reported. Also, scientists found a variety of rodents, porcupines, two species of hyena, a wolf, a fox, a saber-toothed tiger, a leopard, a brown bear, a cave bear, two horned rhinoceros, a woolly rhinoceros, big-horned sheep, a musk ox, a pig, a hare, a deer, a horse camel, bovids, cynailurus, elephants, a baboon, an ostrich, etc.

Shockingly, these findings represent an extraordinary variety of climates and geographical conditions. Investigators agree that the animals found represented European habitats and jungles, warm-moist climates, tundras, and cold-wet geographical zones. Similarly, there were animals of prairies and steppes, dry-temperate climate, deserts, or hot-dry climates. In fact, in one cave at Choukoutien in China investigators found large numbers of complete skeletons of normally incompatible animals like hyena, horse, red deer, tiger, etc. Certainly, it seems that these animals have been interred in the flesh! Furthermore, scientists were shocked to discover at the cave at Choukoutien the fractured remains of seven human beings squeezed into a fissure or narrow cave, representing European, Melanesian, and Skimo racial types. The scientist that found these human remains asked the question of how or why such diverse racial types congregated simultaneously, and how is it possible that seven human beings of different and far-apart geographical zones found themselves far removed from their respective habitats with animals of different geographical zones jammed into fissures? Obviously, floods, hurricanes, winds, rain, snow, earthquakes, and tornadoes do not pick and choose or select its victims. We see in these events a great level of selectivity, which nature by itself does not have. Certainly, it is impossible to think that a great flood removed the European, Melanesian, and Skimo racial types by pure accident and jammed them into fissures. Indeed, it is beyond my imagination to think of what kind of mechanism acted in a worldwide systematic form.

I think the intelligence behind the UFO phenomenon uses a very advance technology of "action at a distance." This idea in physics is that one body can affect another without any intervening mechanical link between them. The use of the term implies a remote and instantaneous influence by the body without any apparent mechanism for transmitting the force produced. I think is possible that the UFO phenomenon uses the same technology to transport boulders, raise the oceans during the Biblical Flood, move fauna and flora from different regions and latitudes, and create mass extinctions.

By the same token, in numerous caves in Australia where there were no ice sheets to blame, in Wellington, a scientist reported findings of animals: "Frequently these occur so fixed between large rocks that it is quite impossible to get them out; and indeed in general none can be got in an entire state from the matrix being in their embedded state full of fractures. The bones were in abundance, and generally upright!"

ESSAY ON THE THEORY OF THE EARTH: ELECTROMAGNETISM IN UFOS AND THE ORIGIN OF MASS EXTINCTIONS AND THE ICE AGES

The animals showed great violence with fractures everywhere; it was clear that it was not a natural death. A scientist said, "They are not found in any regular position, such as would be imagined had their owners lived and died where their remains now lie. Heads, jaw, bones, teeth, ribs and femurs are all jumbled and concreted together without reference to parts. The quantity of small animals it must have taken to form a deep deposit of their bones, perhaps two feet deep, ten wide and of indeterminate length must have been something prodigious, for they are compressed into the smallest possible space." Some parts of the skeleton were embedded in the cement when the ligaments still bound the bones together! The scientists that studied these caves stated, "It is difficult to imagine how that could take place under any natural process with which we are acquainted." Moreover, single bones have been found wedged between huge rocks with the breccia-like mortar between them.

Now assuming that a great flood took place, since when did the transparent water have material properties to jam material objects between and in fissures? Certainly now, in the twenty-first century, the water does not have these type of properties. Also, scientists stated that "including some apparently dismembered while still in the flesh were entombed chaotically and violently and crowded unnaturally into small rock cavities and crevices." So the question is, since when the water has the power to dismember animals and jam them into fissures?

Furthermore, in Maryland, in the United States, there is a famous cave named the Cumberland Cave, where scientists have found animals from different regions and habitats. Brother G. Nicholas writes, "In this one cave have been found such types as the Wolverine, Grizzly Bear and Mustelidae which are native to Arctic regions! Peccaries, the most numerous type represented, Tapirs and an Antelope possibly related to the present day Eland are indigenous to tropical regions! Ground Hogs, Rabbits, Coyotes, and Hare remains are indicative of dry prairies, but on the hand such water-loving animals as beaver and Muskrat suggest a more humid region!" The question is, what made rabbits run into the same cave with carnivorous-like coyotes and antelope with a wolverine and a grizzly bear? Also, bones of mastodons and reptiles were found.

Moreover, J. W. Gidley of the United States Museum, the first scientist to study the cave, stated that "the bones were so hopelessly intermingled that the only reasonable interpretation consistent with the evidence was that the animals were living contemporaneously and were deposited at one time. This refutes the theory that the Northern species were deposited during a Glacial period, and the Southern species during an interglacial Period."

Obviously, it is impossible for nature to operate in this fashion. We see that even in normal emigration patterns, birds do not intermix with different species, the same with deer, bears, etc. So how can we explain this phenomenon? The only one I can think of is that the intelligence behind the UFO phenomenon is responsible for the mass extinctions. The intelligence behind the UFO phenomenon has the technology to create these unique conditions. I think they use some kind of action at a distance, which is a "remote and instantaneous influence by a body, without any apparent mechanism for transmitting the force produced." Also, they use teleportation, which is the theoretical transportation of matter through space by converting it into energy and then reconverting it at the terminal point. Furthermore, I think the intelligence behind the UFO phenomenon has the secrets of entanglement, which is "a bizarre shifting of physical characteristics between nature's tiniest particles no matter how apart they are." Entanglement lies at the heart of teleportation.

In addition, the UFO phenomenon is omnidirectional, which is able to receive or send radiations equally well in all directions. Also, it is omnificent or unlimited in creating power. Similarly, the UFO phenomenon is omnifarious, which has a great diversity of forms or kinds. By the same token, the UFO phenomenon is omnipotent, which is the state of being with force of unlimited power. Also, the UFO phenomenon is omnipresent, which means that it is present in all places at all times! And I think the UFO phenomenon is omniscient, having infinity awareness, understanding, and insight, and possessed of universal or complete knowledge.

Certainly, these are not divine qualities, and I would like to warn the believers that what we have is a very advanced civilization, probably from another dimension and with millions of years ahead of humankind. Further, it seems to me that the intelligence behind the UFO phenomenon manipulates space and time.

Likewise, a scientist, A. H. Verril, in his book Strange Story of the Earth writes, "In an arid valley in Chihuahua Mexico, paleontologists found thousands of skeletons. The place looked as if cordwood had been scattered over it. Giant bison, mastodon, camels, six horned antelopes, horses, rabbits and gophers, besides carnivorous, wolves, saber-toothed tigers, hyenas, badgers and bears."

By the same token, Alcide d'Orbigny, a friend of Charles Darwin, wanted to explain the thousands of horses, mastodons, armadillos, the glyptodont, and the Megatherium buried in the Papampian graveyards of South America. He writes, "I argue that this destruction was caused by an invasion of the continent by water. The sudden movement of the sea, all at once the continent and caused the violent floods of water which carried off the soil, was the result of the sudden rise of the cordilleras." As a matter of fact, Sir Henry Howorth writes, "The sudden rise of the Andes is said to have shaken the whole world with a mighty earthquake, hundreds if not thousands of cubic miles of the body of the earth almost instantaneously heaved upward, produced a violent earthquake which spread throughout the entire globe." The Rockies, the Andes, and a "third the circumference of the globe undergoing simultaneous orogenic movements, with deluging waves sent careening over the land."

Now my question is, what is the connection between the end of the Ice Ages and the rise of the Andes, Rockies, and Himalayas? What does the former have to do with the latter? Obviously, the end of the Ice Ages triggers other phenomena, which have nothing to do with the Ice Ages. So what kind of forces operate deep in the center of the earth?

Furthermore, in another animal graveyard in Palermo, Sicily, in Italy, a scientist found "20 tons of hippopotamus bones dug out of the top of a hill. The bones are those of successive generations of hippopotamus that went there to die. The bones are those of all ages down to the fetus and it's hard to think of the baby hippos accompanying their parents up the slope and lying down to die beside them." Furthermore, scientists reported that "bones of rhinoceroses, camels, giant wild boars . . . It is estimated that the bones of about 9,000 complete animals have been dug out of one hill at Agate Springs, Nebraska. What could have killed so many animals all at one spot?"

Scientists said also they found seashells, fish, plants, and trees "all mixed in great confusion, in great heaps, large and small animals, herbivorous and carnivorous, mammals and birds all in one pile in alluvial deposits." After all, Charles Darwin said, "All attempted have failed to solve the mystery of extinctions and fossil graveyards."

Sir Henry Howorth writes, "Darwin like so many others, who have looked the problem face to face, confessed to me that itremained to him the one stupendous mystery, in the latter geological history of the world for which no rational explanation had been forthcoming."

Similarly, geologist James N. Zumberg of the University of Michigan stated, "We are still unable to account for the loss of the gigantic animals of the Pleistocene. A few geologists are willing to admit quite frankly that the riddle of the mammoths is insoluble conundrum. How the mammoths were enabled to exist in a region where their remains became so speedily frozen? and how such vast quantities of them became accumulated at certain spots, are questions that do not at present seem capable of being satisfactorily answered and their discussion would accordingly be useless."

Another scientist, Charles H. Sternberg, writes, "What caused the death of the countless individuals in the Sternberg Quarry is a question not easily answered. Some scientists believed that during the Upper Miocene Period there were many watercourses separated by slightly elevated divides and broad floodplains and during a rainy season of unusual duration, the whole region for many miles must have been converted into a series of lakes. and all the animals in the vicinity, after having gathered at the highest points, they could not escape death, must have been overwhelmed by some great flood that covered every inch of ground Western Kansas is one vast cemetery."

Daly writes, "What caused reindeer, bear, wolves and mammoths to leave their bones on the top of the Montagne De Sautenay in southern France? How can the mountain of bones be explained on the island of Cerigo near Crete? it is a mile in circumference at the base, and from the base to the summit is covered with bones." Obviously, a flood would not pile a mountain of bones, but it would disperse in an extensive area. Furthermore, at the Cumberland Cave, which I mentioned before, scientists found a great diversity of animals from different geographical zones. "Wolverine, lemming, the long-tailed shrew, mink, red squirrel, muskrat, porcupine, hare, and elk" intermix with animals of warmer climatic conditions like peccary, crocodile, and tapir. "Also animals that now lived on the western coast of America like coyote, badger and a puma-like cat."

Further, there are animals of humid, water like climates like beaver, muskrat, and mink intermixed with animals of arid regions like coyote, badger, and those of wooded regions intermixed with animals of open terrain, like the horse and the hare. Scientists stated, "May it be concluded, that the transportation of these living organized bodies, if such a thing ever happened, has taken place from north to south or from east to west; or was it effected by means that irregularly stirred and mingled them together?"

Furthermore, in a cave in Tasmania, investigators found bones of four hundred opossums and two thousand different kind of mice, bats, porcupines, and small birds; in total, the remains of not less than 6,888.500 individual opossums, mice, and porcupines lie entombed in the cave. Now the question is, why did the Stone Age man waste so much time killing and buried mice and porcupines when there were millions of megafauna waiting for them to be eaten? In addition, there were thousands of bones of birds, lizards, frogs and also large animals like the Megatherium, Mylodon, Toxodon, Macrauchenia, giant armadillo, and jaguars. Also there were bones of mammoths although mammoths are seldom found in caverns.

In one cave, specially around the margin of Lagoa do Sumidouro near Santa Lucia in Portugal, investigators found human bones intermixed with animal bones of horses, llamas, capybaras, etc. The bones of over fifty human beings of both sexes and every age from infant to decrepit old man. Their skeletons lay buried in hard clay and were discovered mixed together in such confusion, not only with each other but with Megatherium bones and those of other Pleistocene animals as to preclude the idea that they had been buried by human agency. Interestingly, all the remains, "both animal and human, possessed the same chemical composition indicating contemporaneity." Moreover, the skulls of these ancient human beings were dolichocephalic (or longheaded), a very interesting feature, because according to Professor Scott Elliot and other scientists, the skulls seemed to be of the old European Neanderthal race, thought to have been ancestral to modern humans and to have become extinct twenty-five thousand years ago.

Similarly, in Gypsum Cave, near Las Vegas, scientists found "unusually well preserved fresh looking remains of the camel, horse and the mountain sheep." As a matter of fact, scientists agreed that "in very recent times great changes in the moulding of the earth surface over a great part of Europe occurred with sufficient rapidityto cause a great destruction of animal life. The tremendous if disorderly concentrations in certain caves and fissures of the bones of just one animal, or very few types are replicated on nearly all continents. Such distribution indicates that the cause of this phenomenon cannot as is often postulated, have resulted, however indirectly, from even the most widespread ice action, that it is possible to image. action which as we have seen in any case almost never occurred." By the same token, in California, scientists found thousands of different species and habitats of birds, and they asked the question, "For what agency could have brought together in one place such dissimilar avian species?"

The scientists found "grebes, herons, bitterns, storks, wood ibises, spoonbills, swans, various geese, snow geese, ducks, American vultures, kites, different kind of hawks, falcons, eagles, caracaras, the Teratornis, quails, cranes, partridges, turkeys, rails, gallinules, parrots, coots, plovers, stilts, sandpipers, barn owls, seven other owl species, flycatchers, woodpeckers, swallows, jays, crows, magpies, titmice,

ESSAY ON THE THEORY OF THE EARTH: ELECTROMAGNETISM IN UFOS AND THE ORIGIN OF MASS EXTINCTIONS AND THE ICE AGES

chickadees, meadowlarks, shrikes, two species of black birds, redwings, orioles, fincasses, sparrows, and buntings." All these bird remains were discovered by scientists in the Late-Pleistocene Tar Pits at McKittrick in California and the asphalt pits at Rancho La Brea in the same state. "This extraordinary assemblage is not an isolated freak occurrence," scientists stated because in late Pleistocene deposits in San Pedro, California, also, an intermix of birds and other animals was found: "camels, bison, groundsloth, rodents, murrolet, black footed albatross, black vented shearwater, fulmar, brand t's, cormorant, green winged teal, mallard duck, cinnamon teal, white fronted goose, surf scooter California quail, turkey vulture and western meadow lark."

The scientists stated, "It is more than probable that most if not all these birds never originally dwelt together in the places where we now find their bones. They were brought together from various directions involuntarily by irresistible winds and buried in a common grave by. It would appear catastrophic agencies, again how does one correlate the extremely varied remains?"

Obviously, you do not need to be a rocket scientist to know that the wind, hurricanes, typhoon, storms, tornadoes, and twisters do not have the intellectual capacity to pick and choose different kind of birds around the world and bury them in a determined place. Since the beginning of history, there is not one event in which a hurricane or typhoon brought together thousands of different animal species and buried them. So how is possible that nature acted different ten thousand years ago?

Furthermore, in the lignite deposit at Geisel valley in Germany, scientists found plants, insects and mammals. They stated, "The amazing preservation of the soft parts of many of these organisms suggest a very re-recent origin of the entire assemblage. The insect fauna for example is a modern one." Although complete insects are a rarity—the majority seem to have been torn apart—the process of fossilization of all surviving parts with "silica invading the tissues must have been virtually instantaneous." Certainly, that is the reason for preserving the membranes and original colors of the insects so fresh.

Furthermore, the most shocking are the plant remains. Investigators reported, "The plant remains are also perplexing, fungi and algae still attached to leaves impressed into the lignite are today found only on plants in Brazil, the Cameroons and Java, chlorophyll is also preserved in many of the leaves which numbering literally billions, form huge beas within these lignites." The scientists proceeded, "The leaves belong to plants from all parts of the world, not just from one or two climatic zones only. The leaves are also mostly shredded so that only their fine fibres or nervous systems remain intact. The fibres often retain their original green (chlorophyllic) colour and indicate that the leaves must have been rapidly excluded from contact with air and light, and buried almost immediately they were stripped from the parent plants."

Again, we find ourselves with the pick-and-choose dilemma—can the wind and hurricane pick and choose plants and leaves all around the world and bury them in a certain place? Obviously, the answer is a loud and clear no! It must be another agency creating the illusion that nature is doing this, and that agency is the intelligence behind the UFO phenomenon. I explained before that the UFO phenomenon uses teleportation, action at a distance, and the entanglement laws of physics in the execution of mass extinctions.

La Brea Asphalt Pits at Los Angeles

3'000.000 bones have been taken from La Brea Asphalt Pits. This is the famous fossil graveyard inside the city limits of Los Angeles. A theory is that herbivores, elephants, camels, deer, horses, bisons, etc., got caught in the oily, sticky mud. After, the carnivores, seeing that easy prey were struggling to get out of the sticky, oily mud, jumped in to eat the prey. So tigers, lions, saber-toothed tigers, etc., followed them in their luck. In Pit no. 3 scientists found seven hundred saber-toothed tigers and seventeen imperial elephants. From Pit no. 9, scientists found camels, horses, ground sloths, mastodons, oak trees, cypress trees, birds larger than condors, coyotes, mountain lions, and human beings.

Another theory about La Brea is that as the centuries passed, like a great procession streaming from surrounding hills and plains, the birds and beasts swept into the blackness of these pools. One by one horses, bison, camels, tigers, lions, saber-toothed tigers, birds, etc., ended in the same sticky death trap. Obviously, we don't see that behavior in animals in Africa of the jungles in the Amazonas. We see animals just like humans; the most important thing is to preserve their lives and survive. So how can we explain this phenomenon? There is no evidence of flood or ice. So it must be some kind of force that caused these animals to jump into the Tar Pit.

Moreover, investigators dug out the mountain lion and tiger cubs, which they showed in every step in the development of the skeleton and dentition. Interestingly, the cubs showed their milk teeth just beginning to appear. Now we know that carnivores nurse their babies in their den. So how is it possible that a baby wolf cub would be out hunting at the Tar Pits? Certainly baby wolves and baby tigers do not jump into the Tar Pit to hunt rabbits. Likewise, the bones of a human being were found mixed with the bones of extinct animals. Certainly, this dates the La Brea Pits during human prehistory, and the skull found is similar to modern humans. Additionally, Carbon-14 dating shows that the trees and extinct animals are dated at 14,000 BP.

Conversely, geologists cannot explain the origin of the Tar Pits because asphalt is an organic residue; where then did the Tar come from that caught the first victim? And why did the first victim not decay if there was no flood mud to cover it? Some scientists postulated the theory that one or two horses and an elephant fell into and created the pit by accident and then a cricket fell and increased the quantity of the Tar Pits. After, a mouse, a grasshopper, added their organic remains until the Tar Pit was in full-scale working condition to trap bison, elephants, tigers, saber-toothed tigers, etc.

My theory is that the intelligence behind the UFO phenomenon is responsible for the tar pits discovered in many parts of the world. Thousands of legends, traditions, and myths mention a calamity that took place within recent human history. For instance, the Greek legends called a visitor that caused destruction in the world Phaeton. Actually what they were describing were UFOs, but they didn't have the technological terms to describe spaceships. So they thought they were stars, moons, and comets creating wonders in the heavens. The traditions and legends mention sticky and inflammable bloodlike fluids falling from the sky as hot naphtha, which is a colorless flammable liquid obtained from crude petroleum, used as a solvent and a raw material for gasoline. Bitumen, which is any of various mixtures of hydrocarbons and other substances occurring naturally or obtained from coal or petroleum found in asphalt and tar. Furthermore, ancient legends and traditions mention rains of fire, which are of hydrocarbon origins. Also, fluids described as bloodlike were not blood but of a brownish red liquid.

For instance, in a legend from Finland, it is called red milk; this implies that it was thick and opaque like the sticky La Brea Tar Pits. Likewise, Greek legends of Phaeton (an assumed star) mention that rains were hail mixed with blood; in fact, it was ice mixed with liquid ferruginous and metalliferous matter. By the same token, the bitter taste of the waters caused by Phaeton were obviously chemical reactions between incompatible substances. Interestingly, a great concentration of ferrous minerals and traces of ferruginous staining exist in various terrestrial surface deposits, obviously laid down by UFOs or Phaeton, as the ancients called it.

For example, bones of Late Pleistocene animals have been found in drift deposits to be stained with iron. Additionally, fossil remains found at high-terrace gravel at Acton and Turnham Green, West London, were found with abundant traces of a Manganese deposit (a brittle gray-white metallic element, symbol Mn, atomic number 25) loaded with manganese oxide blue-gray iron sands. Still, the question is, why and how did thousands of animals in the last Pleistocene mass extinction ended up trapped in the Tar Pits? We know that humans and animals have something in common, and that is the instinct or will to live and survive; that is why it doesn't make sense to think that thousands of animals jumped to their deaths.

ESSAY ON THE THEORY OF THE EARTH: ELECTROMAGNETISM IN UFOS AND THE ORIGIN OF MASS EXTINCTIONS AND THE ICE AGES

My theory is that the intelligence behind the UFO phenomenon used a telepathy-magnetic sense-electroreceptor metathetic stimulus. Biologists and naturalists know that animals in general have a capacity to detect the earth's magnetic field, which acts as an aid to navigation and in their yearly emigrations. For instance, the navigation of European robins can be altered by superimposing an artificial magnetic field. Similarly, pigeons with magnets attached to their heads became disoriented. It's not complicated! The same technique the UFO phenomenon used to telepathy-magnetic affect the megafauna, their magnetic receptors, and made them jump into the Tar Pit. Impossible! Not at all—the phenomenon already happened a couple of centuries ago, but people thought that it was an act of god and called it a miracle.

The story is that St. Anthony of Padua (1231) was at Rimini, Italy, where a great number of atheists congregated. The saint wanted to save the poor souls for the church. So the saint preached to them to have faith in Jesus. Unfortunately, the infidels would not listen. So the saint went to the bank of the river close to the sea, and standing, he began to speak and gave a sermon: "In the name of God unto the fishes hear the word of god, ye fishes of the sea and of the stream, since heretics and infidels are loat to listen to it; and having uttered these words suddenly there came toward him so great a multitude of fishes—great, small and middle sized. As had never been seen in that sea or in that stream or of the people round about; and all held [the fishes] their head up out of the water, and all turned attentively and the greatest peace and Meekness and order prevailed; insomuch that next the stood the lesser fish, and still after them where the water was deepest, stood the larger fish, the fish being thus ranged in order."

St. Anthony began solemnly to preach. Upon these and other familiar words and his teachings, the fishes began to open their mouths and bow their heads! Similarly, in another event, a saint named St. Francis Xavier (1552) was a great missionary who founded many churches with supernatural powers that foretold the future and healed sick people and also amazingly raised several people from the dead! In this particular event, St. Francis was sailing from Ambionum, a city around the Moluca islands, to Baranula when he was overtaken by a storm that threatened to wreck the vessel in which he sailed. All on board pleaded with him to pray for their safety. So the saint took his crucifix against the raging winds, and the storm immediately calmed. But a sudden lurch of the vessel snatched the crucifix from his hand and tossed into the sea. Ultimately, the ship arrived safely the next day at Baranula. As the saint disembarked and was walking along the seashore, according to dozens of witnesses, "A great crab leaped out of the sea carrying the crucifix devoutly and in an upright direction between its claws. The crab made its way directly to the saint, delivered the crucifix to him and then returned to the sea." In these two events, the UFO phenomenon created religion as the Tar Pits created mass extinction.

Furthermore, in Agate Spring Quarry in Nebraska in Sioux County on the south side of the Niobrara River in Agate Spring Quarry is a fossil-bearing deposit up to twenty inches thick. In this quarry, paleontologists have found bones of mammals like twin-horned rhinoceros and a horse named Moropus with legs and claws like those of carnivores, also a giant swine that stood six feet high. In addition, investigators found skeletons of an animal similar to a gazelle and a camel and was called a gazelle-camel. A scientist named A. L. Kroeber got to the conclusion that some of the animals were not more than three thousand years old!

The scholar William R. Corliss described another bone bed in Florida. Many scientists remarked that is one of the richest fossil deposits in the United States. "It has yielded the bones of more than 70 species of animals, birds and aquatic creatures. About 80 percent of the bones belong to Plain animals such as camels, horses, mammoths, bears, wolves, large cats and a bird with an estimated 30-foot wing span. Shockingly, mix with all these animal investigators found shark's teeth, turtle shells, bones of fresh salt water fish. The bones are all smashed and jumbled together as if by some catastrophe." The big question is, how did bones from such different ecological niches like plains, forests, and oceans come together at the same place? I explained before, the UFO phenomenon uses teleportation, action at a distance, and the physics laws of entanglement to create this mixtures of species of different ecological habitats and mass extinctions. The reason that I postulated the theory that the intelligence behind the UFO phenomenon is responsible for mass extinctions and the Ice Ages is because this is the only extraterrestrial agency to operate in the planet after seeing that during mass extinctions and the Ice Ages, unique factors work in the making of upheavals unparalleled in human history.

On Mass Extinctions or Slaughter?

I was in a field, a vast, grassless, sad field. It did not seem to be day or night. I was walking with my brother, the brother of my childhood, this brother of whom I must I admit I never think and whom I scarcely remember. We were talking, and we met others who were walking, we were speaking of a former neighbor, who, because she lived by the street, always worked with her window open. Even while we talked, we felt cold because of that open window. There were no trees in the field. We saw a man passing nearby. He was entirely naked, ashen, colored, riding a horse the color of earth. The man was hairless, we saw his skull and the veins in his skull. He was holding a stick that was limber, like a twig of grape vine and heavy as iron. This horseman passed by and said nothing. My brother said to me, "Let's take the deserted road," there was a narrow, deep-cut road where we saw not a bush or even a sprig of moss. All was earth colored, even the sky. A few steps farther, and no one answered me when I spoke. I noticed that my brother was no longer with me. I entered a village that I saw. I thought that it must be Romainville [why Romainville?]. The first street I entered was deserted. I turned into a second street at the corner of the two streets a man was standing against the wall. I said to him, "What is this place? Where am I?" He did not answer. I saw the open door of a house, I went in. The first room was deserted. I went into the second behind the door of this room a man was standing against the wall. I asked him "Whose house is this? Where am I?" The man did not answer. The house had a garden. I went out of the house and into the garden. The garden was deserted. I found a man standing behind the first tree. I said to this man "What is this garden? Where am I?" the man did not answer. I wondered through the village, and I realized it was a city. all the streets were deserted; all the doors were open. No living being was going by in the streets or moving in the rooms or walking in the gardens.

But behind every turn of a wall, behind every door, behind everything, there was a man standing in silence. Only one could ever be seen at a time. These men looked at me as I passed by. I left the city and began to walk in the fields, after a while, I turned and I saw a great crowd following me. I recognized all the men I had seen in the city. Their heads were strange, they did not seem to be hurrying and yet they walked faster than I, they made no sound as they walked. Suddenly this crowd came up and surrounded me, their faces were earth colored. Then the first one I had seen and questioned as I entered the city said to me "where are you going? don't you know you've been dead for a long time?" I opened my mouth to answer and I realized no one was there.

—Les Misérables, Victor Hugo (1802-1885)

According to scientists, there have been five mass extinctions that we know in the history of the earth. The first known mass extinction is believed to have taken place in the Ordovician epoch or in the Paleoroicera Era 505 to 440 million years ago. The second mass extinction is believed to have taken place in the Devonian epoch 410 to 360 million years ago. The third mass extinction is believed to have taken place in the Permian epoch 286 to 245 million years ago. Moreover, the fourth mass extinction is believed to have taken place in the Triassic epoch between 245 to 208 million years ago. Further, the fifth mass extinction is believed to have taken place between 146 to 65 in the Cretaceous epoch and is the most famous because of the extinction of the dinosaurs. Interestingly, the last mass extinction took place only 11,000 to 18,000 years ago!

Even so, that this mass extinction took place in the Quaternary Era in recent Pleistocene times is not in text books of geology. I think because the conditions in which they took place were not geologically and scientifically explainable. For instance, the frozen mammoths, the intermix of different geographical zones' fauna and flora, the acceleration of the processes of sedimentation, etc. In the last Pleistocene mass extinction, millions of fauna and flora were lost around the world, and their niches or vacuums never were filled by other fauna or flora. Certainly, those are not the normal conditions under which nature works—that is the reason I have no other agency to suspect for this slaughter apart from the UFO phenomenon. So according to my research, there have been six known mass extinctions, but the seventh mass extinction began with the ancient Roman Empire. As a matter of fact, during the inauguration of the Colosseum, nine thousand animals were sacrificed. The Romans were responsible for the disappearance of lions native to Asia and Europe, and the games continued for seven hundred years. Moreover, now with the industrialization and human greed of society, probably a million species are vanishing every year and increasing global warming. Furthermore, there have been many theories about mass extinctions, but none of them offer a good answer. Certainly, I don't claim to have found out the truth, but my purpose is to bring the discussion and debate to the table and keep looking for the answers. Obviously, to some people, my contribution to the subject probably is not relevant. However, I always like try to discover what the meaning of things is and why things happen. I hope the UFO phenomenon has nothing to do with mass extinctions, because if they are responsible, definitely humankind would be in a very precarious situation. After all, the last Pleistocene mass extinction eighteen to eleven thousand years ago affected every continent on the planet, and this one was the last mass extinction.

Mass Extinction in the Subcontinent of India

The late scholar Immanuel Velikovsky wrote about a mass extinction in the subcontinent of India during the last Pleistocene mass extinction. He wrote, "The great massif of the Himalayas rose to its present height in the age of modern, actually historical man. The highest mountains are also the youngest with their top most peaks.

The mountains have shattered the entire scheme of the geology of the 'long, long ago.'"

The Siwalik Hills are in the foothills of the Himalayas north of Delhi; they extended for several hundred miles and are two to three thousand feet high. In the nineteenth century, their unusually rich fossils drew the attention of scientists. Animal bones of species and genre, living and extinct, were found there in most amazing profusion. Some of the animals looked as though nature had conducted an abortive experiment with them and had discarded the species as not fit for life. The carapace of a tortoise twenty feet long was found there.

How could such an animal have moved on hilly terrain? The Elephas ganesa, an elephant species found in the Siwalik Hills, had tusks about fourteen feet long and over three feet in circumference. A scientist wondered, "It is a mystery how these animals ever carried them, owing to their enormous size and leverage." The Siwalik fossil beds are stocked with animals of so many and such varied species that the animal world of today seems impoverished by comparison. "It looks as though all these animals invade the world at one time. This sudden bursting on the stage of such a varied population of herbivores, carnivores, rodents and primates. The highest order of mammals, must be regarded as a most remarkable instance of rapid evolution of species." Another scientist said, "The hippopotamus, which generally is a climatically specialized type, pigs, rhinoceroses, apes, oxen, filled the interior of the hills almost to bursting."

The great A. R. Wallace, who discovered with Darwin the theory of evolution and natural selection, was shocked and astonished by the magnitude of the Siwalik mass extinction. Many of the genres represented are from hundreds of species extinguished to the last one; some are still represented now in days, but only a few species. Furthermore, scientists agreed that of nearly thirty species of elephants found in the Siwalik Hills, only one species survived in India. The sudden and extensive extinction of the Siwalik mammals is the most shocking and without explanation for geologists and biologists alike. The great carnivores, the varied species of elephants belonging to no less than twenty-five to thirty species. A great number of hoofed animals that found a suitable climate lived in the Siwalik jungles. Some scientists thought that the Ice Age had caused the extinction, but there was no Ice Age in Pleistocene times in India. Other scientists thought that the Siwalik deposits were made by floods, but this explanation "does not appear to be tenable on the ground of the remarkable homogeneity that the deposits possess and a uniformity of lithologic composition in a multitude of isolated basins at considerable distance from one another." The scientists agreed that "there must have been some agent that carried these animals and deposited them at the feet of the Himalayas." Moreover, geologists said, "If the cause of these paroxysms and destruction was not local, it must have produced similar effects at the other end of the Himalayas and beyond that range. Thirteen hundred miles from the Siwalik Hills, in Central Burma deposits of fossils of mastodon, hippopotamus, ox is similar to the deposits of Siwalik Hills. Also is found fossil-wood associated with them, hundreds and thousands of entire trunks of silicified trees and huge logs suggested the denudation of thickly forested areas. So, animals met death and extinction by the elementary forces of nature which also uprooted forests and from Kashmir to Indo-China threw sand over species and genre in mountains thousands of feet high."

Again we see that there is no clear explanation about the sudden extinction of so many species of animals and how they ended up in the hilltop of the Himalayas. So the question is, what caused the sudden extinction of thousands of species? We see that there no evidence of ice sheets or a great flood to transport the fauna uphill to the foot of the Himalayas.

Mass Extinction in the Subcontinent of Australia

Similarly, the mass extinction in Australia was total, with the most diverse fauna disappearing. Naturalists have long lists of quadrupeds that vanish forever, for instance, seven-foot-tall kangaroos, rhino-size browsers, giant flightless birds, millions of small mammals, birds, and even small rodents. Scientists still ask if humans or the Ice Age were the cause of their demise. Among the carnivorous predators, the marsupial lion, the largest mammalian carnivore, weighed up to 350 pounds and up to 30 inches tall at the shoulder. Also there were leopards and wild dogs that stalked the megafauna. Furthermore, there is the Victoria Fossil cave, which is a warehouse for bones of more than forty-five thousand animals.

According to scientists, some of the oldest bones belonged to animals larger and more frightening than today we find in Australia. The giant megafauna of the Pleistocene epoch was like a giant wombat the size of a rhinoceros. Also, naturalists have found a tapir-like creature, a hippo-like beast, and a lizard twenty feet long that ambushed its prey and swallowed it completely. The Australian megafauna dominated the continent, and then suddenly, every animal that weighed over one hundred pounds went extinct. What caused the extinction of millions of animals? Furthermore, in the Kelly Hill caves on Kangaroo Island, scientists have found thousands of remains of animals. One of the theories about the Australian mass extinction is the Blitzkrieg Hypothesis, which the paleontologist Paul Martin postulated, which is that humans destroyed very animal species as they spread through the North American continent. Also, this theory is known as the Ruthless Man Theory.

The theory of the Ruthless Man postulated that a group of Stone Age men was armed with stones, spears, probably axes, and sometimes, fire. The Stone Age man, without GPS, hunting gear, rifles, horses, all-terrain vehicles, helicopters, infrared binoculars, or all-weather boots extinguished in every continent—with the exception of Antarctica—animals from mice to birds and mammoths. In my opinion, it is an ingenuous theory without any fundament in reality. According to scientists, what happened to Australia's megafauna "is one of the planet's most baffling paleontological mysteries." Furthermore, scientists blamed the extinctions on climate change. Likewise, the Australian paleontologist Tim Flannery postulated another theory: "Humans who arrived on the continent fifty thousand years ago, use fire to hunt, which led to deforestation and a dramatic disruption of the hydrologic cycle." Flannery said, "Something dramatic happened to Australia's dominant land creatures."

Interestingly now in Australia, the largest land animal is the red kangaroo. The irony is that the modern natives of Australia, even with the boomerang, have not been able to wipe out the kangaroo, which they have been hunting for thousands of years! So how is it possible that eleven thousand years ago, the natives of Australia were more able hunters than modern natives? Undoubtedly, there is no evidence that humans killed the megafauna in Australia. Indeed, it will be helpful if someone discovered a giant wombat with a spear in the ribs or a giant bird in a human campfire, although no discoveries of such kind have been made.

One famous critic of Flannery's overkill theory is Stephen Wroe of the University of South Wales, who wrote, "If this were a murder trial it would be laughed out of court. How could primitive hunters with only spears and fire have eradicated so many species? Moreover, the population it was small now they can cover such a vast continent like Australia with different habitats and regions. Extinction by definition it means that there can be no survivor, but always there is selectivity."

Mass Extinction in South America

Charles Darwin was shocked when he saw the graveyards of the megafauna left by the Pleistocene mass extinction in South America. He wrote under January 9, 1834, in the journal of his voyage, "It is impossible on the changed state of the American continent without the deepest astonishment, formerly it must have swarmed with great monsters. Now we find mere pigmies, compared with the antecedent allied races. The greater number, if not all of these extinct quadrupeds lived at late period and were the contemporaries of most of the existing sea-shells. Since they lived no very great change in the form of the land can have taken place. What, then has exterminated so many species and whole genera; the mind at first is irresistibly hurried into the belief of some great catastrophe; but thus to destroy animals, both large and small, in Southern Patagonia, in Brazil, on the cordillera of Peru, in North America up to Behring's straits. We must shake the entire frame work of the globe. No lesser physical event could have brought about this wholesale destruction, not only in the Americas, but in the entire world."

Interestingly, the discoveries of vast quantities of animal remains in almost every country of South America have been made in recent geological formations. In South America the Pleistocene beds are developed on a very large scale. Similarly, as in Europe and North America, scientists have found caverns of Pleistocene times, especially in Brazil by professors Lund, Claussen, Branard, and Liais. Likewise, rich deposits of Pleistocene age have been discovered in Patagonia, Argentine. Furthermore, in Tiahunaco in Peru, radiocarbon dating of materials indicated that the site is much younger than expected. The early classic stile was dated to about the time of Christ. According to geologists, the city of Tiahunaco continued to be occupied as late as the eighth century AD. According to geologists, it would seem from these dates that a geological catastrophe took place. This estimated Humboldt found more fossil bones in the Cordillera of Chiquitos near Santa Cruz de la Sierra and elephant teeth near Conception in Chile. The most remarkable finding by Alexander Humboldt was on the volcano Imbabura in Quito, Ecuador at an elevation of 7,200 feet above sea level.

Similar in 1795 was the discovery of the skeleton of a Megatherium in Argentine. Interestingly, with this discovery, science has been introduced to a new and old world that had existed in South America in very recent geological times, and humankind has been contemporaneous with it. Additionally, after the great paleontologist and scholar George Cuvier's work was presented to the great geologist Charles Lyell, it was shown in the Museum of the American Philosophical at Philadelphia—a block of limestone from Santas in Brazil obtained by Captain Elliot of the US navy in 1827. "The block contained some human skull teeth, and other bones, together with fragments of shells, some of which still retained traces of their original colors. Remains of several hundred other human skeletons were dug out of similar calcareous Tufa at the same place, where the presence of Serpulae in the rock suggested that all the remains were deposited through marine action." Moreover, he observed, "The shells would not have been brought so far inland by natives for food." Likewise, a great scholar, Dr. C. D. Meigs, who wrote about this discovery, said, "Captain Elliot, while riding along the banks of the river Santas on his way to the town of St. Paul, found a mound 3 acres in extent and 14 feet high, about 10 miles from the sea. The bones he took with him to America, were dug from the face of the hill were it was cut by the wash of the stream, and are parts of one skeleton out of many hundreds that are still lying in their bed of tufa. They were lying on the rock in an oblique direction, the heads uppermost and the lower extremities dipping at an angle of from 20 to 25 below the horizon. Portions of the bones were invested externally with a stalactitic deposit of carbonate of lime, looking very much like a mummified skin. Close to one of the teeth was a serpula and a piece of oyster-shell. The rock in which the skeleton was embedded consisted of fragments of shells united by a stalactitic matter and contained nodules of carbonaceous matter." A question asked by the scientists was about the date of the catastrophe, which enclosed several hundred human beings in that tufa of the Rio Santas. The scientists agreed that "It seems unlikely that these remains were formally buried by sorrowing friends." Also they stated, "It is unlikely that so solid a stone should have been formed at so great a distance from the sea. No doubt they are coexistent with the emerged land; they are not to be considered as the results of human industry."

The shore of the Atlantic must have formerly swept nearly in a line with these remarkable deposits within this bed, or nearest than it to the sea, are found fossil bones of elephants, which cannot be so old as the unfossilized oyster shells since they could not have been fossilized anterior to the existence of the soil out of which they were dug. Furthermore, in a limestone cavern on the borders of the Lagoa do Sumidouro in Santa Lucia in Brazil, the late scientist and scholar Dr. P. W. Lund excavated the bones of more than thirty human skeletons of both sexes and different ages. "The skeletons lay buried in hard clay overlying the original red soil forming the floor of the cave and were found mixed together in such great confusion, not only with one another but with the remains of a megatherium and other Pleistocene mammals. As to preclude the idea that they had been entombed by the hand of man. All the bones, whether human or animal, showed evidence of having been contemporary with one another." Therefore, the mass extinction in South America was total with humans and animals being destroyed in an unimaginable catastrophe created by the Intelligence behind the UFO phenomenon.

Mass Extinction in the North American Continent

The best way to introduce the mass extinction that took place in the North American continent is with the words of the late scientist, scholar, and Professor Frank C. Hibben. He stated that more than forty million animals vanished. He wrote,

> The Pleistocene period ended in death. This was not ordinary extinction of a vague geological period which fizzled to an uncertain end. This death was catastrophic and all inclusive. The large animals that had given the name to the period became extinct. Their death marked the end of an era. But how did they die? What caused the extinction of 40 million animals? . . . the [extinction] was of such colossal proportions as to be staggering to contemplate . . . The "corpus delicti" . . . may be found almost anywhere . . . the animals of the period wandered into every corner of the New World not actually covered by the ice sheets. Their bones lie bleaching in the sands of Florida and in the gravels of New Jersey.
>
> They weather out of the dry terraces of Texas and protruded from the sticky ooze of the tar pits of Wilshire Boulevard in Los Angeles. Thousands of these remains have been encountered in Mexico and even in South America. The bodies lie as articulated skeletons revealed by dust storms, or as isolated bones and fragments in ditches or canals. The bodies of the victims are everywhere in evidence. In

the great deposits of Nebraska, we find literally thousands of these remains together. The young lie with the old, foal with dam and calf with cow. Whole herds of animals were apparently killed together, overcome by some common power . . . . the muck pits of Alaska are filled with evidences of many thousands of animals killed in their prime. The best evidence that we could have that this Pleistocene death was not simply a case of a Bison and the Mammoth dying after their normal span of years found in the Alaskan muck. In this dark gray frozen stuff is preserved, quite commonly, fragments of ligaments, skin, hair, and even flesh. We have gained from the muck pits of the Yukon Valley, a picture of quick extinction. Neither the Pleistocene animals nor they untimely end are phenomena to the American continents. Asia was deeply involved.

Obviously, this was not a natural mass extinction, but a slaughter! Who to blame? the Ice Ages? or a band of "ruthless man" without horses, rifles, shoes, GPS, all-terrain vehicles, infra-red equipment, helicopters, satellites, etc. So how did they manage to kill every mouse and mammoth in the North American continent, at least forty million animals in the United States alone? Also, more millions in Canada. All mammoths, mastodons, horses, tapirs, camels, ground sloths, glyptodonts, peccaries, cheetah, saber-tooth cats, bears, rodents, deer, muskoxen, moose, and millions of birds. Also, millions more animals perished in Canada. Certainly, if modern American Indians could not wipe out the bison without the help of the white man, the horse, and the rifle, how could the Stone Age man or a group of ruthless men have wiped out more than forty million animals without counting the millions in Canada? Furthermore, we can't blame the Ice Ages because the ice was in retreat and the weather was no different than the climate. On the contrary, the climate was much better for the surviving of millions of animals. Again we have no options; if nature or man didn't cause the mass extinctions, something or someone caused them! My only and prime suspect is the intelligence behind the UFO phenomenon. I have explained at the beginning of the book the probability that the UFO phenomenon caused mass extinction.

ROBERT ITURRALDE

Mass Extinction in Europe, Africa, and Asia

The mass extinction in Europe was similar to the extinction in the American continent. For instance, the rhinoceroses (woolly and hairless species), mastodons, woolly mammoths, giant deer, hippos, giant rats, musk-oxen, hyenas, antelope elephants, and millions of animals. According to some studies, 75 percent of all animals over twenty pounds.

Furthermore, scientific studies have shown that 90 percent of all large mammals or seventy genera were exterminated from the continents of Europe and Asia. Although Africa's mass extinction was less severe than the mass extinctions in Australia, South America, the North American continent, and Europe. Moreover, the scientist from the nineteenth century, Henry Hoyle Howorth, in his masterpiece The Mammoth and the Flood, wrote about the extinction of the bird moa in New Zealand. He wrote, "Two researchers Dr. Haast and Dr. Moore in 1868 found a site. The complete sets of leg bone he had examined from this site made up 144 adults and 27 young birds. Similarly, a deposit of Moa bones was found in the winter of 1881 near Motanau. They formed a thickly compacted bed from a few to eighteen or twenty inches in thickness. a second deposit of bones was found at Hamilton in a pit about thirty yards from the first pit. Fully one third of the bones were those of nemiornis and dinornis chiefly of the smaller species and young Moas. The occurrence of these masses of bones, consisting of so many species of different habitats, of wingless geese at home on the water, and of a gigantic raptorial bird with great powers of flight, of gigantic moas and small kiwis, all unrolled and unbroken, suggest similar conclusions which have already forced themselves upon us elsewhere."

In regard to their deposition by human hands, Mr. Booth (a scientist) says, "I must confess that when first commencing to open up the pit. I could see no other cause for the deposit of the bones than that savages had placed them there. But during further progress ofthe work. I was involuntary obliged to abandon my favorite theory. Not the slightest indication of human agency could be detected during the whole course of exhumation. If these bones had passed through the hands of savages, the rude stone implements used by them in those early periods, for the cutting of flesh and the breaking of bones would have left some marks! Among so many hundred bones, some of them would have borne the marks of a sharp edge "a hack or scratch, or a fracture. Some of them would have been split or chipped, or even broken to pieces, for the marrow, that is if they had marrow in their bones and as they were a flightless and even wingless bird. I believed it would be difficult to prove that their bones did not at least contain a little oily matter, which would be a sufficient inducement for the savage to break them or suck them out. But to grant that they had none does not alter the force of the argument. The pelvis is so peculiarly formed on the inside, with covered in hollows, that each one must have contained enough succulent matter to fed a half a dozen savages. Although, I, could adduce reasons for supposing that they might throw large bones into the water hole. Still! cannot see what would cause them to gather up bones about their camp, carry them to a water hole and throw them in. No implements, weapons, toy trinkets, or relic of the bone of a man or a dog was present. If we cannot attribute their deposition to human agency, we cannot invoke carnivorous animals, there were none in New Zealand, besides the bones are ungnawed and sharp. Nor can we attribute the collection of these remains in one spot to the raptorial birds, for their remains are found alongside of those of the moa, apparently sharing in a common destruction."

Again we are facing another mystery like the extinction of the mammoth and other fauna in the Pleistocene. In this case there is no evidence that man was the agent of destruction or the Ice Ages. Only the UFO phenomenon can cause such a perfect slaughter. Likewise, scientists agreed that neither nonexistent northern ice sheets nor southern nonglacial causes are valid scenarios, and some alternative agency responsible for creating the recorded effects must be brought to light. By the same token, the fact is that fauna and flora have been dispersed so violently that sufficient time has not elapsed for new species to evolve.

The evidence of the fossils reveals that that extinction was worldwide, and many species perished literally where they stood, being suddenly buried erect or even in walking positions. Also, like the mammoths, fauna and flora were flash frozen with exceptional speed following their destruction. Moreover, the deposition of many enveloping deposits' drift was catastrophically sudden and extremely fast. In fact, this is not a normal work of nature; this evidence points to a supernatural agency. In this case, the only one I know is the intelligence behind the UFO phenomenon. Some scientists declared, "The vast quantities of vegetable matter represents whole forests which have been obliterated and buried catastrophically. When the mammoth and rhinoceros lived in northern Siberia, these desolated islands were covered with great forests and bore a luxuriant vegetation."

Furthermore, in Europe immense herds of animals vanished off the face of the earth. For not obvious biological or physical reasons, they were well adapted to their natural environment. Likewise, the same biological demise took place simultaneously in Australia, Asia, South America, North America, and Africa. Professor Pilgrim stated, "At approximately the same time we witness a similar extinction of the mammal faunas of Africa and Asia, though in their case, this may not have been caused by glacial conditions." What conditions could these have been? Conditions acting as effectively outside the alleged limits of the hypothetical polar ice sheets as within them and on a hemispheric scale? Furthermore, the deposition deep inland of millions of contemporary marine shells and the stranding at great elevations of marine mammals such as whales, walruses, and seals.

Additionally, vast forests were flattened and buried under vast accumulation of sand or mud in broken and twisted form.

In addition, the form in which this extinction took form and everyday experience that we know. For instance, on incrustation that I mentioned before at the beginning of this chapter, the great scientist, paleontologist George Cuvier wrote, "It is also possible that the animals inhabiting shells may leave their stony coverings when they die in some particular places and that these, cemented together by slime of greater or less consistence, or by some other means, may form extensive banks of shells. But we have no evidence that the sea or ice has now the power of agglutinating these shells by such a compact paste or indurated cement, as that found in marbles and calcareous sand stones or even in the coarse limestone strata in which shells are found enveloped. Still less do we find the sea making any depositions at all of the more solid and silicious strata which have preceded the formation of the strata containing shells."

The Effects of UFOs in Nature

A witness reported a UFO sighting in his garden. The next day, he went to check, and to his amazement, the peach tree was dead! The branches and twigs were shriveled grotesquely, the leaves curled and crisp brown, and the once-healthy buds of peaches looked like prunes. A little digging showed that the tree was killed to its very tap roots—overnight! The tap root is the main part of a plant growing straight downward from the stem. Moreover, other effects of UFOs on nature have been reported by witnesses. For example, trees burning black, tree branches dried and curling, trees carbonized as if petrified or turned to stone, grayish stain on greenery, grass scorched, shrubbery set on fire, radioactivity level in patches of vegetation. Sometimes the plants and vegetation grow faster than normal.

Moreover, in January 1995 in a farm in the United States, a couple of farmers reported UFO sightings. One of the farmers went early to check his livestock and found forty chickens literally frozen to death. In fact, it seems that the animals didn't panic; they were not scattered, just in their coops, dead. Also, three sheep were dead and had been shaved around the cheekbones, and they had a hole drilled into the cheek, all the way into the bone itself! No blood was found! The farmers stated, "We cut the veins of the sheep, and there was no blood, not even in the ground. Strange tracks were found near the farm. Nobody could identify the tracks." The farmer looked at the dog, and his eyes were closed. The farmer opened his eyelids and there were no eyeballs.

Fifteen minutes later, the dog was dead. The farmer stated, "Saturday morning we went to feed the chickens, the sheep, and the dog as usual. It was a quiet morning, and there had had been no unusual sounds or even barking in the yard. At 6:30 a.m., when we woke up, it was already too late. All the animals were already dead."

Likewise, the strange story of a ship named the Orange Medan, which was found wallowing in the Indian Ocean in 1948 with its entire crew dead and their faces frozen in a contortion of horror.

Cattle Mutilation and Mass Extinctions

Cattle mutilation is another aspect of the UFO phenomenon that has been around in modern times for at least two hundred years that we know of. The first case of cattle mutilation that we know took place in 1897 in Leroy, Texas. The witness was Alexander Hamilton, a member of the US House of Representatives, who stated, "We were awakened by a noise among the cattle, I arose thinking my bulldog was performing some of his pranks, but upon going to the door, saw to my utter astonishment an air ship slowly descending upon my cow lot about 40 rods from the house. Calling my tenant, Gid Heslip, and my son Wall, we seized some axes and ran to the corral. Meanwhile the ship had been descending until it was no more than 30 feet above the ground and we came within 50 yards of it. It consisted of a great cigar-shaped portion, possibly 300 feet long, with a carriage underneath. The carriage was made of glass or some other transparent substance alternating with a narrow strip of some material. It was brilliantly lighted within and everything was plainly visible. It was occupied by six of the strangest beings I ever saw. They were jabbering together, but we could not understand a word they said. Every part of the vessel which was not transparent was of a dark reddish color. We stood mute with wonder and fright, when some noise attracted their attention, and they turned the light directly upon us. Immediately on catching sight of us, they turned on some unknown power, and a great turbine wheel about 30 feet in diameter, which was slowly revolving below the craft began to buzz and the vessel rose lightly as a bird. When 300 feet above us, it seemed to pause and hover directly over a two-year-old heifer, which was bawling and jumping, apparently fast in the fence. Going to her, we found a cable about half inch in thickness made of some rare material fastened in a slipknot around her neck, one end passing up to the vessel, and the heifer tangled in wire fence. We tried to get off, but could not, so we cut the wire loose and stood in amazement to see the ship, heifer and all, rise slowly, disappearing in the northwest. We went home, but I was so frightened I could not sleep. Rising early Tuesday, I started out by horse hoping to find some trace of my cow. This I failed to do, but coming back in the evening I found that Link Thomas, about three or four miles west of Leroy, had found the hide and legs in his field that day. He thinking someone had butchered a stolen beast, had brought the hide to town for identification, but was greatly mystified at not being able to find any tracks in the soft ground. After I identified the hide as my brand, I went home, but every time I wanted to sleep, I would see the cursed thing, with its big light and hideous people. I don't know whether the devils or angels, or what, but we all saw them, and my whole family saw the ship and I don't want any more to do with them."

Furthermore, as a member of the House of Representatives and other high social-political circles, Alexander Hamilton staked his honor upon the truth of the story in an affidavit signed for the most important people of his town. Likewise, on September 9, 1967, near the King Ranch in the San Luis Valley, Colorado, a horse named Snippy didn't show up, so the owners started looking for him. The owners found him a quarter of a mile from the ranch. The horse was dead, and from the neck up, it was without flesh. The rest of the carcass was intact, and there was no blood for evidence. The area to the front of the shoulders was also without flesh. It looked like it had been cut with a surgeon's scalpel. Using test areas, the investigators found radiation beyond normal counts.

As in the Hamilton case in 1897, there were no tracks of any kind within one hundred feet of Snippy carcass. Bushes 150 feet from the horse appeared to have been crushed by something coming from above. Interestingly, in the two days, neither insects nor predators had touched the carcass, a shocking similarity with thousands of mammoths that have been found since the nineteenth century. By the same token, in the 1970s, cattle mutilation was in the newspapers in Minnesota, Wisconsin, Kansas, Nebraska, Iowa, South Dakota, Colorado, Texas, Arizona, and California. The cattle mutilation was scientifically done like a surgeon. The ears were removed, tongues were cut out, udders, sex organs, and anuses were sliced out with surgical skill and the blood drained. Again, no footprints or vehicles tracks were found near the carcass. Also, farmers before find their cattle mutilated, they report UFO sightings. Similarly, on July 6, 1975, at the entrance of a North American Air Defense Command (NORAD) installations near an unmarked road that doesn't show on maps. The entire area is a military installation where every few feet there is a warning that "Violators will be prosecuted to the extent of the law." Nonetheless, cows were found dead without sex organs, and many of the carcasses seemed to have been dropped from the air. They found a bull with all four legs broken. Other animals were found in a pasture with a padlocked gate!

In Colorado, there were cattle mutilations near very sensitive military installations. NORAD in Cheyenne Mountain is the agency's operation center. Its installations are a monumental work of engineering; forty-five thousand cubic yards of rock to be blasted out to accommodate its building. Some of the buildings are three stories tall. The purpose of this fortress is to detect and analyze anything that enters or crosses United States airspace. Shockingly, on Tuesday, October 21, 1975, a 1,500-pound female buffalo carcass was found mutilated without udder or ears, and the vagina had beenremoved. This incident raised eyebrows because if the most advanced human technology can't detect UFOs and its occupants because they became invisible, it means that they can be anywhere and everywhere without detection. Furthermore, on October 17, 2001, eight cows were reported mutilated. A farmer said, "I have lived in this country all my life and worked on ranches and seen plenty of dead animals, but never did I see an animal with its face mask removed like that and the reproductive organs also removed."

The Intermix of Fauna and Flora

Obviously, the ice, floods, mud, and water have the property of mixing different flora and fauna from different geographical zones. Interestingly, this phenomenon was not an exception but the rule during the last Pleistocene mass extinction. The great geologist and scholar R. F. Flint wrote, "In Cromer, Norfolk, close to the North Sea coast and in other places of the British Isles, forests beds have been found. The name derives from the presence of a great number of stumps of trees once supposed to have rooted and grow where they are now found. Many of the stumps are in upright positions and their roots are often interlocked. Today these forests are recognized as having drifted: The roots do not end in small fibers but are broken off, in most cases one to three feet from the trunk. Bones of sixty species of mammals, besides birds, frogs and snakes, were found in the forest bed of Norfolk.

"Among the mammals were the saber-toothed tiger, bear, mammoth, straight tusked elephant, hippopotamus, rhinoceros, bison, and modern horse. two exclusively northern species—glutton and musk—ox were found among animals from temperate and tropical latitudes. Of the thirty species of large land animals of the forest bed, only six still exist in any part of the world. All the others are extinct and only three are presently native to the British Isles. Remains of sixty-eight species of plants were obtained from the Norfolk forest bed; They indicate a climate and geographical conditions very similar to those of Norfolk at the present day. In view of the sensitivity of plants to thermal conditions, the conclusion might well be drawn that the climate at the time the forest-bed was deposited was not different from the present, which conclusion the fauna, comprising southern as well as northern animals contradicts. The abundance of animals of so many different species on an island the size of Great Britain caused speculation that one time it must have been part of a continent and that the strait of Dover was not then opened. It is necessary to point out, however that the opening of the straits of Dover is a geological revolution of considerable magnitude, such as one might well hesitate to ascribe to the comparatively short period embraced by glacial and post-glacial time."

Furthermore, above the forest bed, there is a freshwater deposit with arctic plants—arctic willow and dwarf birch and land shells. Geologists agree that is a remarkable change from the climatic conditions of the forest bed below, which indicates a lowering of temperature of about twenty degrees. The geologists ask, "Whatcould have brought, together or in a quick succession, all these animals and plants, from the tundra of the arctic circle and from the jungle of the tropics, from lush oak forests and from desert, from lands of many latitudes and altitudes, from freshwater lakes and rivers and from the salt seas of the north and the south?"

Furthermore, the shells with closed valves show evidence that the mollusks did not die a natural death but were buried alive. It seems that this agglomeration was brought together by a moving force that rushed over the continent and swept marine animals and deep-water creatures and swept the British Isles with animals from the tropics. Undoubtedly, we know that nature doesn't work in this manner. However, the question is why the intelligence behind the UFO phenomenon caused this kind of mass extinction and turned the planet upside down. I think the idea is to confuse us; that is the reason millions of animals have been deposited in fissures, crevices, and caves around the world. Ask this question: how will the world look with millions of animal carcasses in the surface of the earth? Also the idea is not to make easy for humankind to answer the question of why, how, and for what reason we are on this planet. I don't see the hand of Allah and Mohammad nor the hand of Jehovah and Jesus nor of any god in the cause of mass extinctions. The only hand that I see causing mass extinctions is the intelligence behind the UFO phenomenon. Another question is why the UFO phenomenon causes mass extinctions. The reason is to allow evolution to proceed in its work and to create intelligent forms of life. Probably there are millions of planets with life that never had the chance to evolve intelligence. For instance, here on Planet Earth, the dinosaurs lived over one hundred million years without developing intelligence.

ESSAY ON THE THEORY OF THE EARTH: ELECTROMAGNETISM IN UFOS AND THE ORIGIN OF MASS EXTINCTIONS AND THE ICE AGES

So it was not an accident that the asteroid or some kind of disaster helped the evolution of mammals to advance to man. The intelligence behind the UFO phenomenon planned. So if you are religious, it's fine to say that it was God who planned the creation of man, is good to live with illusions. I think this is the only reason for humankind to unite and fight for the survive of the human race because probably the UFO phenomenon will set the clock of destruction with the destiny of Israel. I think this is the ground zero, and the UFO phenomenon will act according to the destiny of Israel.

In summary, mass extinctions are supernatural and extraterrestrial; there is nothing on this earth that can wipe out life on this planet. The late great scholar Immanuel Velikovsky wrote, "But even a sudden climatic catastrophe all over the world could hardly have been adequate by itself to account for an extermination so wide and, for many species so complete. Climatic change alone is not enough to explain the extinction of the marvelous Pleistocene fauna. There have been other suggestions, such as clouds of volcanic gases, which destroyed whole herds of mammals. Of what dimensions must these clouds have been? They must have covered all the terrestrial globe. But all the volcanoes of the earth, erupting together would not be sufficient to destroy so many species and genera."

I understand that is not part of everyday routine to explain the supernatural or UFOs. But for me, the best example through which I can explain the form that mass extinction took place during the last Pleistocene mass extinction is to compare it to the Shroud of Turin. The reason is that these supernatural phenomena took place in a form that only a global, omnipresent power like the UFO phenomenon can cause. For instance, Carbon-14 tests of the Shroud of Turin give the date from AD 1260 to AD 1340. Furthermore, analysis of pollen from the shroud shows evidence of pollen from the famous cedar of Lebanon; this is evidence of the shroud of having being in Provenance, Palestine. Also analysis gave evidence of grains from typical halophytes, plants which are common to the desert regions around the Jordan Valley. These types of plants are unique, because they are adapted to live in a soil with a high content of sodium. Chloride is found almost exclusively around the Dead Sea. Among these were desert varieties of Tamarix, Suaeda, and Artemisia.

Dr. Frei, the scientist that analyzed the shroud, said, "These plants are of great diagnostic value for our geographical studies as identical desert plants are missing in all the other countries where the shroud is believed to have been exposed to the open air. "Consequently, a forgery produced somewhere in France during the Middle Ages in a country lacking these typical Halophytes, could not contain such characteristic pollen grains from the desert regions of Palestine." D. R. Frei believed that the pollen collected from the shroud includes six species of exclusively Palestinian plants and what he describes as a significant number of plants from Turkey, mostly from the Anatolian steppe. In addition, there are eight species of Mediterranean plants. So it seems that the shroud sometime in its history was exposed to the air in Palestine and Turkey. Moreover, the overall style of the weave is like most Palestinians, Romans, and Egyptians and the weave is possibly from the first century. In addition, D. R. Frei found traces of cotton. The fibers correspond to the species Gossypium herbaceum which is characteristic of the Middle East. Although it is significant that cotton has been found at all. Surely, its presence determines conclusively that the fabric of the shroud came from the Middle East, since cotton is not grown in Europe. So the Carbon-14 test of the shroud revealed to have been made between AD 1260 to AD 1340, but the other tests revealed to components from the Middle East, which is impossible for anyone to do, because how can you travel in time to the first century and collect pollen to make the shroud? I believed only the UFO phenomenon is able to do this miracle. Similarly, scientists have found that fauna and flora of different geographical zones intermix, which nature cannot create. Only the UFO phenomenon can create a worldwide destruction like the last Pleistocene mass extinction.

CHAPTER 6

The Origin of the Ice Ages and the UFO Phenomenon

The development of these huge ice sheets must have led to the destruction of all organic life at the earth's surface. The ground of Europe, previously covered with tropical vegetation and inhabited by herds of great elephants, enormous hippopotami, and gigantic carnivora became suddenly buried under a vast expanse of ice coverings plains, lakes, seas and plateaus alike. The silence of death followed . . . springs dried up, streams ceased to flow, and sunrays raising over that frozen shore . . . were met only by the whistling of northern winds and the rumbling of the crevasses as they opened across the surface of that huge ocean of ice.

—Louis Agassiz (1807-1873)

ow in the twenty-first century, what caused the Ice Ages is still a mystery. There are at least sixty theories; among them are the following:

Geophysical Theories

1. The Continental Drift theory postulates the slow drifting of crustal continents toward or away from each other. Polar

wandering or slow changes in the earth's axis of rotation, crustal sliding.
2. Land-water changes
Oceanic changes, relocated oceans rearranged oceanic circulation or changed directions of ocean currents.
3. Land mass altitude changes
Larger land areas brought above the permanent snow line, feeding the growth of ice sheets.
4. Glaciological theories
Periodic melting of basal ice layers of natural ice caps, causing massive surging of ice leading to formation of huge peripheral ice shelves, which in turn displace seawater to create sudden rise of the world's oceans.
5. Atmospheric changes
Variations in carbon dioxide content in the atmosphere, reduction of amount of CO_2 and reduced rate of absorption by the oceans, variations in the volumes of volcanic dust and gas, variations of ozone content.
6. Axial and orbital changes
Alterations in axial tilt of the earth, leading to loss of terrestrial light changes in the intensity of composition of solar emissions received by the earth.
7. The galactic dust theory
Furthermore, in 1997, Richard A. Muller of the University of California, Berkeley, and Gordon J. Macdonald theorized that the earth passes through a particular band of cosmic dust every

100.000 thousand years. Moreover, in the 1920s, the Yugoslavian mathematician Milutin Milankovitch postulated a theory that now is called the Milankovitch Cycles.

The theory stated that are three kinds of variations that affect the orbit of the planet in its movement around the solar system. First the earth travels in an elliptical orbit, not in a circular form but in an oval shape like an egg. Although, there is an eccentricity in this

ESSAY ON THE THEORY OF THE EARTH: ELECTROMAGNETISM IN UFOS AND THE ORIGIN OF MASS EXTINCTIONS AND THE ICE AGES

elliptical orbit. In the course of 100.000-year cycles, the orbit becomes less elliptical and more circular and then moves back to the elliptical orbit. Second, the earth is tilted, and the angle of the tilt change in the course of a 41.000-thousand-year cycle, from a maximum of 24.5 degrees away from the vertical to a minimum of 21.5 degrees. Third, the earth spins around its axis like a top with a wobble in it. The wobble is called precession, and it keeps a 22.000-thousand-year cycle. Also, an additional small skip appears every 19.000 years. He got to the conclusion that at the extreme end of the precession cycle and the tilt cycle, the amount of solar energy diminishes enough to allow ice to accumulate and create the Ice Ages. The scientific establishment accepts this theory as the best possible explanation of the cause of the Ice Ages although there are many perplexing climate behaviors that the theory can't explain. For instance, from 1400 to 1800, it was so cold that those years were called the Little Ice Ages. As a matter of fact, global temperatures dropped by a degree or two. Interestingly, the Dutch canals froze over daily and the Swedish army invaded Denmark by marching across the iced North Sea. On the other hand, scientists don't know why the period between 1100 and 1250 was so warm in Europe and America, and shockingly the Vikings grew crops in Greenland. Obviously, the Milankovitch cycles can't explain these sudden fluctuations in the temperature of the world. Furthermore, according to geologists in the previous 90.000 thousand years, Greenland had more than twenty abrupt temperature fluctuations. The changes suddenly increased the snowfall, the amount of dust in the air changed by a factor of ten, and average temperature rose or dropped by twenty degrees. Actually, these changes occurred quickly, in decades and sometimes in as little as a few years. In fact, the bigger shifts lasted for about a thousand years and were preceded by smaller back-and-forth jumps. Certainly, this abrupt change of temperature variations can't be explained by the Milankovitch cycles or meteors hitting the earth or volcanic eruptions because they are too infrequent. In addition, there was a cold snap a thousand years long that ended eleven thousand years ago, while the earth was emerging from the Ice Age and entering a warm phase.

Conversely, this transitional millennium is called Younger Dryas after a little flower that spread rapidly, and the European forests became tundra, the Gobi became a desert, and the oceans shifted.

Baron Alexander von Humbolt, in 1852, pointed that one hemisphere is heating up while the other was cooling down; this was incorrect. He said that the average temperature of either hemisphere is controlled not by the number of hours of daylight and darkness but by the total number of calories of solar energy received each year. Furthermore, d'Alembert's calculation had demonstrated that any decrease in solar heating that occurs during one season because the earth is farther from the sun is exactly balanced by an increase during the opposite season. Also, when the earth is closer to the sun, the total amount of heat received by one hemisphere during the year is always the same as that received by the other hemisphere. Moreover, Sir Fred Hoyle wrote, "The Milankovitch model shows that cyclic variations in the earth's temperature are to be expected. It does nothing to show that ice ages would not occur even if all these variations were absent." Mr. Hoyle didn't believe that the size of the cycles "was sufficient to plunge the earth into an ice age or pull it out of one." In addition, he said, "If I were to assert that a glacial condition could be induced in a room liberally supplied during winter with charged night-storage heaters simply by taking an ice cube into the room, the proposition would be no more unlikely than the Milankovitch theory." Similarly, he believed that the slow astronomical changes could never account for the suddenness of the onset of the Ice Ages. Shockingly, warm periods have been about 10.000 years long, while the Ice Ages have been about 100.000 thousand years long. Moreover, the Milankovitch cycles do not explain simultaneous Ice Ages in the north and south hemispheres. Also, there is evidence that ice sheets existed in India, Africa, and tropical zones.

The great geologist A. P. Coleman mentions the "insoluble mystery of an Icecap in the hot tropics on the 2.000 feet high plateau of southern India." Furthermore, geologists discovered the so-called "glacial till" and "striae," which they interpreted as a Permian Ice Age in the hot tropics. Although the great scientist Ernst J. Opik admits, "How this could happen in a region which at present is within the tropics, stretching from 17 to 24 north latitude, is one of the greatest geological puzzles we are confronted with."

Moreover, just as Africa and India developed ice caps, Antarctica is believed to have developed tropical coal swamps. The late geologist A. P. Coleman produced enough evidence that a continental ice sheet or an ice cap invaded Africa from the sea. Furthermore, in India, the evidence is that the ice moved from the south toward the north in a direction away from the equator as far as the Salt Range of Pakistan. Certainly, this is an insoluble enigma for modern geology, for ice from the equator flowing northward cannot be fitted into any of its theories. Furthermore, most Ice Age theories cannot explain the great amounts of water necessary to form the ice caps, because cold weather means less evaporation, and less evaporation means less snow to pack as ice caps. Similarly, no Ice Age theory can explain how to make the earth cold and hot at the same time—cold enough to freeze the ice caps but warm enough to evaporate the large amounts of water needed to form the ice caps. Moreover, Louis Agassiz and other Ice Age theorists cannot explain how the ice developed before the rise of the great elevations like the Alps, Himalayas, Andes, Rocky Mountains, and other great elevations. Modern geologists agreed that high mountains are necessary to provide and replenish the snow from which glacier ice is derived to produce an Ice Age. Interestingly, most of the world's present major ranges attained their present elevations approximately 11.000 thousand years ago. So the question is, where was the elevations necessary to attract the heavy snowfalls to form the ice caps before 11.000 thousand years ago? Similarly, there are more questions—for instance, how is water transported from temperate and equatorial regions to polar regions in order to increase the ice caps? So if many of today's highest mountains were much lower when the Ice Age was at its zenith, how did so much ice accumulate to form the continental ice caps?

By the same token, it takes a lot of heat to make an ice cap; water must be evaporated in great quantities to fall as snow in the polar regions. Shockingly, the same heat would then prevent the necessary cooling of the ice caps! Conversely, a good Ice Age theory must provide a good explanation for all the phenomenon that accompanies the beginning and ending of an Ice Age. For instance, (1) an initiating event or condition, (2) a mechanism for cyclic repetitions or oscillations during the period of glaciations, (3) an ending event or condition, and (4) it must solve the mystery of increased precipitation with the increase of cold.

Furthermore, geologists thought that in glacial periods the ice caps would advance from the poles. Shockingly, now it seems that the ice sheets started from different centers and expanding in all directions. So the poles have nothing to do with the expansion and increasing of the ice sheets. Moreover, some of the coldest places on earth like Siberia didn't develop ice sheets. Furthermore, according to some geologists, the ice caps were suddenly born. In fact, the Wisconsin ice sheet did not start in a small area and expand outward, but rather, it started all at once over a great area. It was fast enough so that the ice cap did not start growing at the coast and move inland; it started in the Hudson Bay region, and after it had grown thick enough to move, it spread outward in all directions. As a matter of fact, the best evidence of the suddenness of the ice cap birth is the discovery that it contained thousands and perhaps millions of animals of a temperate climate, many of them frozen entirely into the ice, including mastodons, mammoths, bear, elk, beaver, etc. When the ice cap melted, many of the animals were dropped into bogs, which preserved their bodies and sometimes the contents of the stomach. The best evidence that the Ice Ages can take place any season of the year and has nothing to do with the winter season is the Berezovka mammoth. The Berezovka mammoth, geologists found, was frozen in the middle of August.

So the Berezovka mammoth is the best evidence that the Ice Ages have nothing to do with the winter seasons or the Milankovitch cycles. In fact, in the mammoth tongue as well, between the teeth, were found portions of the last animal meal. Shockingly, it seems that it had no time to swallow! I explained in the previous chapters about the Berezovka mammoth.

As a matter of fact, the Berezovka mammoth was found in the muck, not in ice, because there never were ice sheets in Siberia! So it means to me that the Ice Age began not with ice and snow but a very cold air that freezes everything before the snow and ice accumulates. Obviously, these statements seem crazy, but we are talking about a supernatural event created by the intelligence behind the UFO phenomenon. The Ice Ages are not a natural phenomenon. Additionally, there is archaeological evidence that there was a human population in northeastern Siberia in Paleolithic times, in Neolithic times, and in the Bronze Age. Likewise, Paleolithic artifacts were found in Yakutia. Archaeologists found rock drawings very similar to the paintings found in the caverns of France and Spain in the valley of Lena not far from the place where the Berezovka mammoth was found near the village Shishkino.

In the Neolithic Age, two or three millennia before the birth of Christ, descendants of earlier inhabitants of Yakutia spread to the coast of the Arctic Ocean in the north and the Kolima in the east. Furthermore, on the lower Lena River, inside the polar circle, archaeologists have found monuments of past cultures. For instance, archaeologists found a metallurgist workshop from the second millennium BPE where they made bronze axes, bronze tips for the spears, knives, and even swords. Undoubtedly, these findings are the evidence of a civilization in the taiga of Northeastern Siberia. Certainly, these findings imply that the climate changed from a temperate climate to a frozen tundra the same moment that the Berezovka mammoth was frozen. Obviously, this is the reason that millions of carcasses of fauna are being found in Siberia, because the climate was ideal for the mammoth, mastodon, and millions of other animals to exist in Siberia. Also, this is the reason archaeologists found evidence of successive groups of human population. So it seems very clear to me that the climate in Siberia was changed suddenly by from beautiful weather to a frozen tundra. Obviously, nature doesn't work like this; this is a supernatural event caused by the

UFO phenomenon. The suddenness of the change cannot be explained by any geological theory.

Furthermore, another important question is, what initiates the glaciations? Similarly, scientists have determined that the heat required to evaporate enough water to form a glacier would be sufficient to fuse and transform into glowing molten liquid, a stream of cast iron five times as heavy as the glacier itself! A great scholar and glaciologist, William Lee Stokes, stated, "Lowering temperatures and increased precipitation are considered to have existed side by side on a worldwide scale and over long periods in apparent defiance of climate logical theory." As a consequence, geologists stated, "In both hemispheres were Pleistocene conditions have been studied, this period appears to have been characterized by increased precipitation as well as lowered temperatures." By the same token, if precipitation were then greater over certain areas of the earth's surface than it is at present, it is implied that in large areas, evaporation was greater than normal. Also, it implies that the climate was warmer than normal. I think this is a shocking conclusion! So, how were Ice Ages created in a subtropical climate? Furthermore, a great glaciologist, G. H. Dury, stated, "It is now generally held that glacial maxima occur simultaneously in the north and south hemispheres. Glaciation was simultaneous in Europe and in North America." Such a statement defies any Ice Age theory. Moreover, radioactive isotope dating proves European, American, and southern hemisphere's Ice Ages were simultaneous! Obviously, this is a shocking statement showing that there is no way to categorize the Ice Ages as a natural phenomenon.

Further, there is the question of the ice movement, because it is well known that ice cannot ascend hills. Also, deposits of gravel and other drift materials sometimes occur only on the northern and northwestern flanks of hills. In many instances, it shows that they have been plastered against the hillsides with great force. By the same token, there are many mysteries in the geological history of the earth without a solution. For instance, fifty-six million years ago—and, it seems, a dozen times in the natural history of the earth—there has been a

ESSAY ON THE THEORY OF THE EARTH: ELECTROMAGNETISM IN UFOS AND THE ORIGIN OF MASS EXTINCTIONS AND THE ICE AGES

sudden massive release of carbon. Scientists estimated that the amount of carbon released during the Paleocene-Eocene Thermal Maximum or PETM was the equivalent of humankind burning all the reserves of coal, oil, and natural gas. The PETM lasted more than 150.000 thousand years until the excess carbon was reabsorbed by the oceans and dissipated in the atmosphere. Interestingly, this excess of carbon brought floods, plagues, and extinctions. Obviously, the question is, what caused the sudden super massive release of carbon into the atmosphere? Humankind was not walking the planet at that time, so from where came all the carbon?

I think the answer is the intelligence behind the UFO phenomenon; they accelerate the natural process or any organic process, even life and death. Obviously, such idea seems crazy. However, the following examples are the proof of what I am talking about. As a matter of fact, all the events that I am going to describe, religious people think that God caused them. Nonetheless, now we know that only the UFO phenomenon can cause them because we have more evidence of the existence of the UFO phenomenon than the existence of God. For instance, the UFO phenomenon appeared in the form of the Madonna in Mexico 1531 to a peasant. The Madonna wanted to show to the people that miracles were real. So the peasant named Diego, while walking, heard somebody calling him. So he kept walking and then found himself face-to-face with a lady of overpowering brilliance and beauty. Her garments shone like the sun, and a violet color permeated the surroundings. So the lady told Juan to go and get flowers in the rocky hill not far from where they stood. So he went and was amazed to see a brilliant profusion of flowers including Castilian, roses *blooming in the rocky and frozen soil*. The soil was frozen, rocky, and the flowers were in full bloom; also it was out of season. According to the peasant, it was impossible for any flowers to grow in a terrain so stony that it only yielded cactus and bushes. Interestingly the flowers glittered with dewdrops and "their delicious fragrance rose like a breath of paradise." Obviously, scientists don't believe these type of supernatural events, and I believe this is the part of Siberia that was changed into a frozen tundra.

Certainly, this supernatural event was caused by the UFO phenomenon using the Madonna as a disguise.

Next, in this supernatural event, the evidence is very clear how, probably, the ice sheets were formed in the Ice Ages although religious people believe that it was God responsible for the supernatural event. In the nineteenth century, in a church in Poland, a group of parishioners wanted to build a church, but to cross to the other side to get the rocks, they needed a bridge. The previous winter was too mild to form ice to form a bridge. So the priest of the church began to pray; according to witnesses, almost immediately, on March 15, 1879, a violent gale began to pile up ice floes until a bridge of thick ice formed from one shore to the other. The next day, on Sunday, the parishioners found that the ice bridge could support the weight of thirty to forty people.

Likewise, the next day, more than 150 horse-drawn sledges began crossing the bridge. Interestingly, by the end of the week, every needed stone to build the parish was in place. After everything, when the ice bridge was no longer needed, a thaw set in the following day just as quickly as it had formed. The ice bridge was gone!

Another interesting event is the story of St. Therese of the Child Jesus, affectionately known as the Little Flower of Jesus. She wanted to receive the habit in a ceremony conducted in the snow.

Since there was no snow, she prayed. The next morning, when she walked outside the church, the garden was white with snow. The only problem is that it was in the month of June! Moreover, in this case, in 1604, St. Seraphin of Monte Granaro wanted to visit the shrine at Loreto. Although the River Potenza stood in the way, and that day, the river was very high. So he prayed very hard, and when he walked out into the river, the water instantly became solid ground. So he was able to walk back and forth across the river. After he had finished his walk, the water turned back into liquid!

Obviously, all these supernatural events mentioned here are not accepted by scientists because they can't be measured or weighed. Nevertheless, they are real as anything we see. So my theory is that extreme natural conditions like the Ice Ages are not part of nature but created by the intelligence behind the UFO phenomenon. I believe that they increase the natural process like sedimentation, rain, snow, wind, and every process in nature. For instance, the best example is the forty days of rain mentioned in the Bible, and the Flood, which I mentioned in the previous chapter. Certainly, there is no way that the earth by itself can produce a flood of such big proportions. Similarly, with the Ice Ages, the earth by itself can't produce a coldness and ice to last 120.000 years! So the UFO phenomenon increases natural process like life, death, rain, snow, thunder, lightning, and other natural phenomena.

CHAPTER 7

The Tragedy of the Human Condition

> As a reliable compass for orientating yourself in life, nothing is more useful than to accustom yourself to regarding this world as a place of atonement, a sort of penal colony. When you have done this you will order your expectations of life according to the nature of things and no longer regard the calamities, sufferings, torments and miseries of life as something irregular and not to be expected but will find them entirely in order, well knowing that each of us is here being punished for his existence and each in his own particular way.
>
> —Arthur Schopenhauer (1788-1860)

Interestingly, during the fifties and sixties among the circles of ufologists, it was mentioned—the Prison Earth Theory. This theory was postulated as early as 1953 by Al Bender's Space Review magazine. Al Bender asked people to send their ideas about UFOs, the universe, etc. One of the people that answered was Harold F., who wrote, "I think the saucers are to keep us imprisoned on earth until we civilize ourselves to the point we no longer pose a threat to them."

Furthermore, the writer of the book UFOs Confidential named George Hunt wrote that the earth is a "prison world" or a "great lunatic asylum." His idea is that man's freedom is an illusion. Also he believed that the earth is controlled by a great force of aliens from the constellation of Orion.

Similarly, the late UFO scholar and military hero Donald Keyhoe, in his book Flying Saucers: Top Secret, mention the idea that the intelligence behind the UFO phenomenon is here to "keep tabs on a colony that was deliberately abandoned." So this idea is that the earth is like a super-immense Alcatraz that serves as an example to other beings in the universe.

Likewise, Jan Hudson in her book Those Sexy Saucer People mentions a similar theory. She believed that UFOs appeared after we detonated the first H-bombs and began space travel. She wrote, "I think the earth is a giant institution in which the human race has been incarcerated for its own good. And every time we start rattling the bars, the keepers come hurrying down to take a look." By the same token, in the 1950s the self-proclaimed UFO abductee George Adamski claimed that a Saturnian during a conversation was told that the earth had been selected centuries ago as the system to which they exile "their troublemakers, they were gathered in ships from the many planets and transported to earth, without equipment or implements of any kind."

Here we were forced to work to gain a place in the creators' colony. Furthermore, the extraterrestrials watch and send messiahs from time to time like Mohammad, Jesus, Buda, Confucius, and Zoroaster. Similarly, a UFO contactee in the 1950s named Orfeo Angelucci revealed that vast numbers of Earthlings are former inhabitants of the world of Lucifer that once existed between Mars and Jupiter. Those responsible for the destruction were deep drowned in time and matter to live in the underworld of illusion that makesup the earth. UFOs still visit our prison world to liberate us spiritually.

I have been studying UFOs since 1987, and my theory is that the intelligence behind the UFO phenomenon owns this planet and perhaps our universe. Also, the earth is like a giant zoo where they conduct experiments of many kinds like religion, war, greed, and destruction. So we are a giant zoo that is supplied and tabbed every day. Another part of the experiment is religion, and Mohammad and Jesus are part of their creation.

Furthermore, there are frightening stories about aliens and UFOs. One story is that they are different kinds of aliens like reptiles, drakes, and Greys living on this planet. One story is the belief that a vast network of alien cities exists; underground cities existed underneath the cities we know. For instance, many ufologists believe that the aliens inhabit large underground bases in the Western United States. For example, the Dulce Base in New Mexico, with its seven sublevels, ufologists believed to be the central facility. From there, the aliens have a transportation system called TUBE shuttles to transport aliens and government officials to sub cities under the following cities: Pago, Arizona, New Mexico, and Colorado Springs

Furthermore, according to UFO researchers, signs on doors, hallways, and tube shuttles are in some type of language. Also, the underground facilities are controlled by some sort of magnetic technology, and they are crowded with reptilian aliens working with government officials. In this facilities, aliens conduct genetic breeding experiments. Nevertheless, the very existence of these facilities is still debatable. Although the base has been mentioned in hundreds of websites, the base has never been found.

Also, among ufologists, there is the story that human-reptoid crossbreeds make up 5 percent of human population. Interestingly, according to a 2002 Roper Poll conducted for the Scifi, now called Syfy channel, one percent of Americans or about three million people claimed to have had close encounters with aliens. Similarly, 5 percent of professional astronomers polled in 1952 claimed they had personally seen UFOs. Another story reported by ufologists is that the reptilian Grey crossbreed alien drinks the blood of animals.

Moreover, ufologists describe the underground of the Dulce Base where experiments are conducted. Level 6 is privately called Nightmare Hall. It holds the genetic labs, where experiments are done on fish, seals, birds, and mice that are vastly altered from their original form. There are multi-armed and multi-legged humans and several

ESSAY ON THE THEORY OF THE EARTH: ELECTROMAGNETISM IN UFOS AND THE ORIGIN OF MASS EXTINCTIONS AND THE ICE AGES

cages of humanoid batlike creatures as tall as seven feet. The aliens teach humans about genetics. Likewise, at Level 7, there are humans in cages, rows after rows of thousands of humans, human-mixture remains, and embryos of humanoids kept in cold storage. Also, the humans in cages are dazed or drugged, but sometimes they cry and beg for help. The aliens say that they are hopelessly insane and involved in high-risk drug tests to cure insanity. Also, the aliens say to never talk to them.

As I have shown there are so many stories said about UFOs and aliens that are hard to believe. After researching UFOs for thirty-one years, the tragedy and paradox of the human condition is thatthe UFO phenomenon is responsible for what we call religion, supernatural phenomena, and mass extinctions. So what we call religion is nothing other than the description of the interaction of aliens and humans since the beginning of history. By the same token, from the bushmen to the Jewish, Christian, and Muslim religions, they are a creation of the UFO phenomenon. Furthermore, everything we call supernatural phenomena like ghosts, hauntings, monsters, near-death experiences, out-of-body experiences, and mass extinctions are a creation of the UFO phenomenon. In fact, I think the UFO phenomenon owns humankind and Planet Earth.

Obviously, it seems preposterous to think like that, but if they gave us religion, then they must have some claim on us. However, now that we know the possibility that extraterrestrials have been involved in our creation and in our future fate, I think it is time to rebel and be independent in the creation of the destiny and fate of humankind. We have the technology and the intelligence to make this world a paradise. We no longer need extraterrestrial help because in these 6.000 years of written human history, we have make such great advances in technology. I think the world has enough resources to feed every human being on this planet and money to make this world better. The only reason we are in this condition is because of human greed, the culture of poverty, and religious fanaticism. I would like to mention an interesting anecdote about a Princeton professor that spoke to Albert Einstein about his son. He was a brilliant college student, but he was very depressed.

He refused to continue his studies or do anything else. So Einstein asked his father what his trouble was. Was he worried about his own death? "No," his father answered. "He is concerned about the death of the solar system. He said, 'Someday it will all go to pieces, and then? Everything accomplished on Planet Earth would go for naught. It would be as if nothing really happened here at all. So why bother to do anything now?'"

Obviously, this is a pessimist outlook of human life. I think still there is room for improvement if we eliminate religious fanaticism and human greed. The job of every government in the world should be the end of poverty. So even knowing that our origins are extraterrestrial—that should not take away the fact that we are probably a unique human species in the universe. Actually, even if the gods of our childhood, Jehovah, Allah, or Khrisna are an illusion created by the UFO phenomenon, human life is still beautiful.

I would like to mention the late Carl Sagan. He wrote, "The earth is a place. It is by no means the only place. It is even a typical place. No planet or star or galaxy can be typical because the cosmos is mostly empty. The only typical place is within the vast, cold, universal vacuum, the everlasting night of intergalactic space, a place so strange and desolate that, by comparison, planets and stars and galaxies seem achingly rare and lonely. If we were randomly inserted into the cosmos the chance that we would find ourselves on or near a planet would be less than 10/33 a one followed by 33 zeroes. In everyday life such odds are called compelling. Worlds are precious."

So this thought should give us the understanding that the human race is unique in the universe and that we should work for the wellbeing of humankind. So if the religious fanatics understood that their gods for whom they are killing are just the product of the human interaction with the UFO phenomenon, we will live in a better world.

Conclusion

> I'm not one of those complicated mixed-up cats. I'm not lookin' for the secret to life or the answer to life. I just go on from day to day, takin' what comes.
> —Frank Sinatra

Yes! Humans can and are causing local extinctions around the world and also are responsible for global warming. For instance, the last pair of auks were killed in 1844 in Eldey near Iceland; the Charles Island Tortoise was slaughtered by overhunting. According to recent surveys, approximately one million species of insects, plants, and animals are lost every year. Similarly, a region the size of England is lost in the Amazon jungle in South America every year. Moreover, it is a fact that since the Industrial Revolution, the earth has been warming up more than normal and the oceans are rising. Although humankind is responsible for global warming and the disappearance of the Amazon jungle with their species of fauna and flora; nonetheless, humankind was not able to cause the last Pleistocene mass extinction. The reasons are very clear, and I explained them in the previous chapters. I think there is no power on earth great enough to wipe out totally every single organism, like we see in all the mass extinctions that we know about. Certainly, the last Pleistocene mass extinction was too strange and peculiar to have been caused by nature. I think mass extinctions are always external in nature; they are a creation of the UFO phenomenon.

The UFO phenomenon intervenes to keep evolution in check and allows the development of intelligence. So the earth and its inhabitants are part of a great experiment in a big lab; also the illusion of religion is part of the experiment.

BIBLIOGRAPHY

Glavin, Terry: *The Six Extinction.*

Gould, Stephen Jay: *Dinosaur in a Haystack*

Rudwick, Martin J. S.: *Georges Cuvier, Fossils Bones and Geological Catastrophes.*

Alvarez, Walter: *T. Rex and the Crater of Doom*

Ryan, William and Walter Pitman: *Noah's Flood*

Simons, Eric: *Darwin Slept Here*

Huxley, Julian: *Evolution in Action*

Burkitt, Miles: *The Old Stone Age Rhodes. Evolution*

Montagu, Ashley: *The Human Revolution*

Greene, John C.: *The Death of Adam*

Howells, William: Back of History

Olson, E. C: *The Evolution of Life*

Haeckel, Ernest: *The Riddle of the Universe*

Kant, Immanuel: *Critique of Judgement*

Kant, Immanuel: *Critique of Pure Reason*

Sagan, Carl: *The Dragons of Eden*

Simpson, George Gaylord: *Life of the Past.*

Gould, Stephen Jay: *Ever Since Darwin.*

Smith, Homer W: *From Fish to Philosopher.*

Iturralde, Robert: *A Treatise on Human Nature: Christian Saints, Historical Figures and the UFO Phenomenon*

Iturralde, Robert: *The UFO phenomenon and the birth of the Jewish Christian and Moslem Religions.*

Hoyle, Henry: *The Mammoth and the Flood.*

Trinkaus, Erik: *The Neanderthals.*

Cro-Magnon: *How the Ice Age Gave Birth to the First Modern Humans*

Coleman, A. P.: *Ice Ages Recent and Ancient*

Raup, David M.: *Extinction: Bad Genes or Bad Luck?*

Lyell, Charles: *Principles of Geology*

Brooks, C. E. P.: *Climate through the Ages*

Fagan, Brian: *The Great Warming*

Strain, Mac B.: *The Earth's Shifting Axis.*

Coyne, Jerry A.: *Why Evolution Is True*

Wilford, John Noble: *The Riddle of the Dinosaur*

Hapgood, Charles: *Path of the Pole*

Colbert, Edwing H.: *Wandering Lands and Animals*

Ward, Peter D.: *The Call of Distant Mammoths*

Gould, Stephen Jay.: *Wonderful Life*

Darwin, Charles.: *The Origin of Species*

Daly, Reginald Aldworth: *The Changing World of the Ice Ages*

Kaufman, Les: *The Last Extinction*

Gribbin, John and Jeremy Cherfas: *The First Chimpanzee*

Also by Robert Iturralde

The UFO Phenomenon and the Birth of the Jewish-Christian and Moslem Religions

A Treatise on Human Nature: Christian Saints, Historical Figures, and the UFO Phenomenon

About the Author

Robert Iturralde has been researching UFOs since 1987. He was a proud member of the US Air Force. By accident, he found a book about UFOs in a flea market. After reading the book he went to the New York Times to check the reports of hundreds of witnesses. The reports were based in real eye - witness sightings. He became very interested in the subject and wrote 4 books about UFOs. Mr. Iturralde in his free time likes to run and play chess.